MATHCO™

COLLEGE ALGEBRA

The *STEP-BY-STEP* Study Guide

L. MURIEL LOCKE

Mathco Publishers LLC
Chesapeake, VA

"The BEST for Your Success"

Mathco College Algebra

Published by Mathco Publishers LLC
www.mathcopublishers.com

Paperback   ISBN-13: 978-1717108074
                    ISBN-10: 1717108075

Kindle e-book   ISBN-13: 978-0985717025

Dedicated to the
Farrar, Jarrett, Locke and Williams families—
past, present and future

CONTENTS

Contents

Chapter 5   Radicals and Equations

Chapter 6   Polynomial Equations and Inequalities

Chapter 7   Logarithmic and Exponential Equations

Chapter 8   Systems of Linear Equations

Chapter 9   Factorial Notation and Applications

Introduction: Math is *Everybody's* Favorite Subject

Welcome to the study of College Algebra.

Mathco College Algebra is an easy-to-follow study guide designed to help you learn and remember many of the important concepts of College Algebra. This notebook contains informative lecture material which you can use in your College Algebra course. It has study tips, helpful hints, informal definitions, and step-by-step worked examples which will assist you in "catching on" to College Algebra. Students in my classes have also used these math notes and examples as a summary when studying for tests. You will find *Mathco College Algebra* to be a resourceful supplement to your class lecture notes and your textbook. Read it and use it to achieve greater success in your mathematics class.

A Guiding Principle: "Learn more and live more."

L. Muriel Locke

About the Author

L. Muriel Locke, an associate professor of mathematics, enjoys helping students to successfully improve their math skills in various math courses including basic algebra, mathematics for the liberal arts, college algebra, precalculus, applied calculus and calculus with analytic geometry. She has a Bachelor of Science degree in Mathematics/Education from Temple University, and a Master of Arts degree in Mathematics from the University of North Carolina at Charlotte. She continued her doctoral studies in Applied Mathematics at Old Dominion University. Professor Locke appreciates the opportunity to share her lecture materials with you, providing clear and understandable presentations of the concepts of college mathematics.

College Algebra Study Tips

- Be sure to attend each class session.

- As an introduction to the Math lesson for the day, read the concepts and examples of *Mathco College Algebra* and your textbook before you go to class.

- Take notes and copy all of the examples worked during the lecture.

- After class, plan to spend <u>AT LEAST ONE HOUR ON EACH PROBLEM SET</u> assigned for homework.

- Review the concepts and examples shown in the *Mathco* study guide thoroughly. Read the class notes and examples carefully.

- Do ALL of the assigned problems and more, not less. Check your answers in the Answer Section of your textbook. If there are errors in the homework problems, make corrections.

- Get extra assistance from your professor during office hours, as needed. Do not wait until the day of a Test to get help.

- When you study for each Test, use your class notes and the *Mathco* study guide to review the rules, definitions, and processes. Then do the Review Exercise assigned by your professor a second time without looking back in your notes or textbook. When you finish, check your answers and make corrections.

- Above all, relax. Keep a positive attitude, and, concentrate on your Math skills.

The Scientific Calculator

There are several brands of scientific calculators and each model may have a different layout for the keypad. In Chapter 7 and Chapter 9 of **Mathco College Algebra**, use the appropriate special keys which appear on your calculator. This applies to both the small-screen calculator and the graphing calculator.

✠ *Helpful Hint* – **The Exponent Key**
On your calculator keypad, you should have ONE of following keys: $[\textbf{Y}^{\textbf{X}}]$ or $[\textbf{X}^{\textbf{Y}}]$ or $[\textbf{^}]$. Whenever the *Mathco* study guide uses the $[\textbf{Y}^{\textbf{X}}]$ key, press the key which is on your calculator, $[\textbf{Y}^{\textbf{X}}]$ or $[\textbf{X}^{\textbf{Y}}]$ or $[\textbf{^}]$.

✠ *Helpful Hint* – **For An Operation Written Above a Calculator Key**
On your calculator keypad, you should have ONE of following keys: **[2nd]** or **[Shift]** or **[INV]**. Whenever the *Mathco* study guide uses the **[2nd]** key, press the key which is on your calculator, **[2nd]** or **[Shift]** or **[INV]**.

PROCEED WITH
OPTIMISM,
CONFIDENCE,
AND PERSISTENCE.

Chapter 1   **Basic Concepts**

1.1   PROPERTIES OF REAL NUMBERS

Let's review some of the basic definitions and properties of College Algebra.

SET NOTATION

A **set** is a collection of numbers or objects enclosed in braces. Let set $A = \{1, 4, 7, 10, 13\}$.
Each number in the set is called an **element** of set A. The symbol $\in$ means "is an element of".
For example, $4 \in A$. (That is, "4 is an element of set A.")

Let set $B = \{1, 7, 13\}$. Then B is called a **subset** of A since set B is part of set A.
The symbol $\subset$ means "is a subset of". In symbols we write $B \subset A$. (i.e. B is a subset of A.)

Union "$\cup$" means combine two sets together. For example, let set $C = \{3, 6, 9, 12, 15\}$ and
set $D = \{2, 4, 6, 8, 10, 12\}$. Then $C \cup D = \{2, 3, 4, 6, 8, 9, 10, 12, 15\}$.

SUBSETS OF THE REAL NUMBER SYSTEM

N = Natural Numbers = $\{1, 2, 3, 4, 5, \ldots\}$
   These are the counting numbers. Natural numbers are positive.

W = Whole Numbers = $\{0, 1, 2, 3, 4, 5, \ldots\}$
   Whole numbers are the counting numbers and zero.

I = Integers = $\{\ldots, -5, -4, -3, -2, -1, 0, 1, 2, 3, 4, 5, \ldots\}$
   The integers are the positive and negative whole numbers and zero.

Q = Rational Numbers
   = {All positive and negative fractions formed by an integer divided by an integer}

   The rational numbers include the common fractions, the integers, the terminating decimals,
   and the repeating decimals.   **EXAMPLES** of Rational Numbers - -

   Common fractions: $\dfrac{2}{3}$; $-\dfrac{5}{4}$ Integers: $3 = \dfrac{3}{1}$; $-5 = \dfrac{-5}{1}$

   Terminating decimals and percents: $0.25 = \dfrac{25}{100} = \dfrac{1}{4}$; $3\% = \dfrac{3}{100}$

   Repeating decimals and percents: $0.\overline{3} = 0.3333\ldots = \dfrac{1}{3}$; $44.\overline{4}\% = 0.\overline{4} = \dfrac{4}{9}$

Observe that the natural numbers, whole numbers, and integers are subsets of the rational numbers.

   (The symbol Q' is pronounced "Q prime".)
Q' = Irrational Numbers = {All positive and negative non-terminating and non-repeating decimals}.
   EXAMPLES   The number $\pi = 3.141592654\ldots$ is an irrational number.
   Radicals (roots) which do not reduce, such as $\sqrt{2} = 1.414213562\ldots$, are irrational numbers.

R = Real Numbers = {Rational Numbers} $\cup$ {Irrational Numbers}. In other words, the **real numbers**
   are all of the positive and negative whole numbers, fractions, decimals, percents, and zero which
   we use for computation. All of the examples used above are real numbers.

1

(1.1)

PROPERTIES OF THE REAL NUMBER SYSTEM

Let A, B, and C represent any three real numbers.

COMMUTATIVE PROPERTY		EXAMPLES
For Addition	$A + B = B + A$	$2 + 5 = 5 + 2$
For Multiplication	$AB = BA$	$2 \cdot 5 = 5 \cdot 2$

For addition and multiplication, we can *change the order* of the numbers and the answer is the same.

ASSOCIATIVE PROPERTY		EXAMPLES
For Addition	$(A + B) + C = A + (B + C)$	$(2 + 3) + 5 = 2 + (3 + 5)$
For Multiplication	$(AB)C = A(BC)$	$(2 \cdot 3)5 = 2(3 \cdot 5)$

To add three numbers, we can *change the grouping* of the numbers and the answer is the same.
To multiply three numbers, we can *change the grouping* of the numbers and the answer is the same.

DISTRIBUTIVE PROPERTY

The Distributive Property is one of the most important and useful skills of College Algebra. The Distributive Property has multiplication outside of the parentheses and addition or subtraction inside the parentheses.

Basic Rule: $A(B + C) = AB + AC$

Multiply the outside term times each term inside the parentheses.

EXAMPLES $6(x + 2) = 6x + 12$ $5(c - 4) = 5c - 20$ $7(3x + 4y - 2) = 21x + 28y - 14$

IDENTITY PROPERTY		EXAMPLES
For Addition	$A + 0 = A$	$5 + 0 = 5$
For Multiplication	$A \cdot 1 = A$	$5 \cdot 1 = 5$

If we *add zero* to a number, the answer is the number. If we *multiply a number times 1*, the answer is the number.

INVERSE PROPERTY	(The "Cancel-out" Property)	EXAMPLES
For Addition	$-A + A = 0$	$-5 + 5 = 0$
For Multiplication	IF $A \neq 0$, $A\left(\dfrac{1}{A}\right) = 1$	$5\left(\dfrac{1}{5}\right) = 1$

If we add a number and its *opposite*, they "cancel-out" to zero. If we multiply a number other than zero times its *reciprocal*, they "cancel-out" to 1.

1.2 LINEAR EQUATIONS IN ONE VARIABLE

An **equation** is a mathematical expression which contains an equal sign. The **root** or solution of an equation is the value of the variable which makes the equation true.

EXAMPLE 1 Show that $x = 3$ is the root of the equation $4x - 5 = 7$.
Answer: Substitute 3 in place of x in the equation to get $4(3) - 5 = 7$, which is true.

When solving an equation, each answer line should have *one equal sign.*

EXAMPLE 2 To solve $\mathbf{2x + 1 = 5}$, is the following statement true or false?

$$2x + 1 = 5 = 2x = 4 = x = 2$$

Answer: This statement is *false* because $5 \neq 4 \neq 2$!

The correct solution is $\mathbf{2x + 1 = 5}$
$2x = 4$ (Subtract 1 from both sides of the equation.)
$x = 2$ (Divide both sides of the equation by two.)

Helpful Hint – The answer lines of an equation should be written downward, not across.

SOLVING LINEAR EQUATIONS

A **linear equation** is a 1st - degree equation. That is, it has no exponents, and, it usually has one root.

Steps:

- *Simplify the parentheses* on the left side of the equal sign.
  Simplify the parentheses on the right side of the equal sign.
- *Combine the similar terms* on the left side of the equal sign.
  Combine the similar terms on the right side of the equal sign.
- *The variable must be on one side* of the equal sign.
  If the variable is on both sides, cancel out one of the variables by adding or subtracting it.
- Now we have a one-step or a two-step equation.
  To solve the equation, *use the opposite of each operation shown in the equation.*

EXAMPLE 3 Solve the equation:

$$3x + 5x - 6x + 5 = 19 - 20$$

$$2x + 5 = -1$$
Combine the similar terms on the left side of the equation, and, combine the similar terms on the right side.

$$-5 \quad -5$$
Subtract 5 from both sides of the equation.

$$2x = -6$$

$$\frac{2x}{2} = \frac{-6}{2}$$
Divide both sides of the equation by 2.

$$x = -3$$
Write the solution.

(1.2)

EXAMPLE 4 Solve: $3(r-2)-(5r-4) = 7-2(4r-1)$

$3r-6-5r+4 = 7-8r+2$	Simplify the parentheses using the Distributive Property.
$-2r-2 = -8r+9$	Combine similar terms on both sides of the equation.
$+8r \quad\quad +8r$	The variable *r* is on both sides of the equation. Add 8r to both sides of the equation.
$6r-2 = 9$	
$+2 \quad +2$	Add 2 to both sides of the equation.
$6r = 11$	
$\frac{6r}{6} = \frac{11}{6}$	Divide both sides by 6.
$r = \frac{11}{6}$	Write the solution.

EQUATIONS WITH FRACTIONS

Helpful Hint – **Eliminate the fractions:** Multiply each term times the lowest common denominator (LCD). For each of the fractions, cancel out the denominator and multiply the result times the numerator. Rewrite the equation and continue the steps for solving the equation.

EXAMPLE 5 Solve: $7y - \frac{5}{3} - 6y = \frac{5}{6} - \frac{1}{3}$

$(6)7y - (6)\frac{5}{3} - (6)6y = (6)\frac{5}{6} - (6)\frac{1}{3}$	Multiply each term by the LCD 6.
$(6)7y - \overset{(2)}{\cancel{(6)}}\frac{5}{\cancel{3}} - (6)6y = \overset{(1)}{\cancel{(6)}}\frac{5}{\cancel{6}} - \overset{(2)}{\cancel{(6)}}\frac{1}{\cancel{3}}$	Cancel out each denominator. Multiply each term of the equation.
$42y - 10 - 36y = 5 - 2$ $6y - 10 = 3$	Combine the similar terms on both sides of the equation.
$+10 \quad +10$ $6y = 13$	Add 10 to both sides of the equation.
$\frac{6y}{6} = \frac{13}{6}$	Divide both sides by 6.
$y = \frac{13}{6}$	Write the solution.

EXAMPLE 6   Solve: $\dfrac{2}{5}(x-3) = \dfrac{1}{4}(5x-4) - x$

$$\frac{2}{5}x - \frac{6}{5} = \frac{5}{4}x - 1 - x$$

Simplify the parentheses using the Distributive Property.

$$(20)\frac{2}{5}x - (20)\frac{6}{5} = (20)\frac{5}{4}x - (20)1 - (20)x$$

Multiply each term by the LCD 20.

$$\overset{(4)}{\cancel{(20)}}\frac{2}{5}x - \overset{(4)}{\cancel{(20)}}\frac{6}{5} = \overset{(5)}{\cancel{(20)}}\frac{5}{4}x - (20)1 - (20)x$$

Cancel out each denominator.

$$8x - 24 = 25x - 20 - 20x$$

Multiply each term of the equation.

$$8x - 24 = 5x - 20$$
$$-5x \qquad\quad -5x$$

Combine similar terms on the right side.
Subtract 5x from each side.

$$3x - 24 = -20$$
$$+24 \qquad +24$$

Add 24 to both sides of the equation.

$$3x = 4$$

$$\frac{3x}{3} = \frac{4}{3}$$

Divide both sides by 3.

$$x = \frac{4}{3}$$

Write the solution.

EQUATIONS WITH DECIMALS

You may solve the equation with the decimals included. Or, you may eliminate the decimals and solve the resulting equation.

EXAMPLE 7   Solve the equation by working with the **decimals included**:

$$0.03(2x - 6) - 0.03 = 0.05 + 0.04x$$

Simplify the parentheses using the Distributive Property.

$$0.06x - 0.18 - 0.03 = 0.05 + 0.04x$$

$$0.06x - 0.21 = 0.05 + 0.04x$$
$$-0.04x \qquad\qquad -0.04x$$

Combine the similar terms on the left side.
Subtract 0.04x from both sides of the equation.

$$0.02x - 0.21 = 0.05$$
$$+0.21 \quad +0.21$$

Add 0.21 to both sides of the equation.

$$0.02x = 0.26$$

$$\frac{0.02x}{0.02} = \frac{0.26}{0.02}$$

Divide both sides by 0.02 .

$$x = 13$$

Write the solution.

(1.2)

A decimal equation can also be solved by **eliminating the decimals**. First simplify the parentheses. If each decimal has one decimal place, multiply the equation by 10. If each decimal has two decimal places, multiply the equation by 100. If each decimal has three decimal places, multiply the equation by 1,000.

EXAMPLE 8 Solve: $0.2 - 0.3(3x + 2) = 0.5$

$0.2 - 0.9x - 0.6 = 0.5$	Simplify the parentheses using the Distributive Property. Each decimal has one decimal place.
$(\mathbf{10})\,0.2 - (\mathbf{10})\,0.9x - (\mathbf{10})\,0.6 = (\mathbf{10})\,0.5$	Multiply each term by 10.
$2 - 9x - 6 = 5$	Simplify each term.
$-9x - 4 = 5$	Combine the similar terms on the left side.
$+4 \quad +4$	Add 4 to both sides of the equation.
$-9x = 9$	
$\frac{-9x}{-9} = \frac{9}{-9}$	Divide both sides by -9.
$x = -1$	Write the solution.

Helpful Hint – Sometimes **the decimal places are "mixed"**. If the maximum number of decimal places is two, multiply the equation times 100. If the maximum is three, multiply the equation times 1,000.

EXAMPLE 9 Solve: $0.03(2x - 5) + 1 = 0.5 - 0.04(1 - x)$

$0.06x - 0.15 + 1 = 0.5 - 0.04 + 0.04x$	Simplify the parentheses. Here the maximum is two decimal places. Multiply each term by 100.

$(\mathbf{100})\,0.06x - (\mathbf{100})\,0.15 + (\mathbf{100})\,1 = (\mathbf{100})\,0.5 - (\mathbf{100})\,0.04 + (\mathbf{100})\,0.04x$

$6x - 15 + 100 = 50 - 4 + 4x$	Simplify each term of the equation.
$6x + 85 = 4x + 46$	Combine similar terms on each side.
$-4x \qquad -4x$	Subtract 4x on both sides of the equation.
$2x + 85 = 46$	
$-85 \quad -85$	Subtract 85 from both sides of the equation.
$2x = -39$	
$\frac{2x}{2} = \frac{-39}{2}$	Divide both sides by 2.
$x = -19.5$	Write the solution.

LINEAR EQUATIONS – TWO SPECIAL CASES

The linear equations we have studied so far are called **conditional equations**. A conditional linear equation has exactly one root or solution. Two special types of linear equations are shown below.

EXAMPLE 10   Solve: $4(x - 1) - x = 1 - 2(x - 5) + 5x$

$4x - 4 - x = 1 - 2x + 10 + 5x$	Simplify the parentheses using the Distributive Property.
$3x - 4 = 3x + 11$	Combine the similar terms on each side.
$-3x \qquad -3x$	Subtract 3x from both sides of the equation.
$-4 = 11$ (False!)	The *x* term canceled out on both sides of the equation and the remaining number statement is false.
No Solution	Write the answer.

This type of equation is called a **contradiction**. It has no solution.

EXAMPLE 11   Solve: $5(2x - 1) - 3x + 2 = 3x + 4(x + 1) - 7$

$10x - 5 - 3x + 2 = 3x + 4x + 4 - 7$	Simplify the parentheses using the Distributive Property.
$7x - 3 = 7x - 3$	Combine the similar terms on each side.
$-7x \qquad -7x$	Subtract 7x from both sides of the equation.
$-3 = -3$ (True)	The *x* term canceled out on both sides of the equation and the remaining number statement is true.
{All Real Numbers}	Write the solution set.

This type of equation is called an **identity**. Observe that the answer line "$7x - 3 = 7x - 3$" has the same expression on both sides of the equation. Since the solution set is {All Real Numbers}, an identity has an infinite number of roots. (This means that you can choose <u>any</u> real number and substitute it in place of x in Example 11. Simplify the equation and the result will always be true.)

RE-WRITING FORMULAS

A **formula** is an equation since it contains an equal sign. **"Solve for y"** means rewrite the formula so that **y** is on one side of the equal sign, and, all other terms and symbols are on the other side of the equal sign.

To solve for the indicated variable, use the opposite of each operation shown in the formula.

EXAMPLE 12   Solve for *y*: $T = wxy$

In this formula, the term wxy is using multiplication. The opposite operation is division.

$\dfrac{T}{wx} = \dfrac{\cancel{wxy}}{\cancel{wx}}$	Divide both sides of the equation by *wx*.
$\dfrac{T}{wx} = y \quad$ or $\quad y = \dfrac{T}{wx}$	Write the answer.

(1.2)

EXAMPLE 13   Solve for **v**: $2u + 3v = 9$

In this formula, the term 2u is being added on the left side of the equation. The opposite operation is subtraction.

$$2u + 3v = 9$$
$$-2u \qquad\quad -2u \qquad\qquad \text{Subtract 2u from both sides of the equation.}$$

$$3v = -2u + 9$$

$$\frac{3v}{3} = \frac{-2u}{3} + \frac{9}{3} \qquad\qquad \text{Divide each term by 3.}$$

$$v = -\frac{2}{3}u + 3 \qquad\qquad \text{Write the answer.}$$

EXAMPLE 14   Solve for **w**: $x = \dfrac{5w + 11}{2w - 3}$

First eliminate the fraction.

$$(2w - 3)\, x = \frac{5w + 11}{2w - 3}\,(2w - 3) \qquad \text{Multiply both sides times the LCD (2w – 3).}$$

$$(2w - 3)\, x = \frac{5w + 11}{\cancel{2w - 3}}\,\cancel{(2w - 3)} \qquad \text{Cancel out the denominator.}$$

$$2wx - 3x = 5w + 11 \qquad \text{Simplify the equation. There are two } w \text{ terms.}$$

Get the *w* terms to the left side of the equation by subtracting 5w to both sides of the equation.

$$-5w \qquad\quad -5w$$

$$2wx - 5w - 3x = 11$$
$$+3x \qquad\quad +3x \qquad\qquad \text{Add 3x to both sides of the equation.}$$

$$2wx - 5w = 3x + 11$$
$$w(2x - 5) = 3x + 11 \qquad \text{Use the Distributive Property (or factoring) on the left side of the equation.}$$

$$\frac{w(2x - 5)}{2x - 5} = \frac{3x + 11}{2x - 5} \qquad \text{Divide both sides by 2x – 5.}$$

$$w = \frac{3x + 11}{2x - 5} \qquad\qquad \text{Write the answer.}$$

EXAMPLE 15   Solve for **F**: $C = \dfrac{5}{9}(F - 32)$

$$\left(\frac{9}{5}\right)C = \left(\frac{9}{5}\right)\frac{5}{9}(F - 32) \qquad \text{Multiply } \frac{9}{5} \text{ on both sides of the equation.}$$

$$\left(\frac{9}{5}\right)C = \left(\frac{\cancel{9}}{\cancel{5}}\right)\frac{\cancel{5}}{\cancel{9}}(F - 32) \qquad \text{Add 32 to both sides.}$$

$$\frac{9}{5}C + 32 = F \quad \text{ or } \quad F = \frac{9}{5}C + 32 \qquad \text{Write the answer.}$$

8

1.3   APPLICATIONS (WORD PROBLEMS)

Steps For Solving A Word Problem:
- Set up a brief *table* or chart: Make a list of the items for which we will find answers.
- One of the items must be identified as *x*. Use the information given in the problem to write an *algebraic expression for each item*.
- Write an *equation* using one variable which is based on the table and the words of the problem.
- Solve the equation. Find *all* of the *answers* requested in the word problem.

GEOMETRY – ANGLES OF A TRIANGLE

The sum of the angles of any triangle is 180°:

$$\angle A + \angle B + \angle C = 180°$$

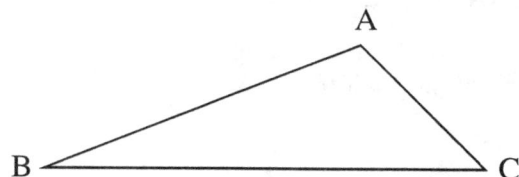

EXAMPLE 1

In a triangle, the measure of $\angle B$ is one-third of $\angle A$ and $\angle C$ is 5 degrees more than $\angle B$. Find the measure of each angle.

✠ *Helpful Hint* – When two items are compared, we usually identify the second item as *x*.

Table	(Set-up)	(Explanation)
$\angle A$	x	$\angle B$ is compared to $\angle A$. Identify $\angle A$ as x.
$\angle B$	$\frac{1}{3}x$	"$\angle B$ is one-third of $\angle A$"
$\angle C$	$\frac{1}{3}x + 5$	"$\angle C$ is 5 degrees more than $\angle B$"

Formula $\angle A\ +\ \ \angle B\ \ +\ \ \angle C\ \ \ \ \ \ =\ \ \ 180°$

Equation $x\ \ +\ \ \frac{1}{3}x\ \ +\ \ \frac{1}{3}x + 5\ \ =\ \ \ 180$

$$(3)x\ +\ (3)\frac{1}{3}x\ +\ (3)\frac{1}{3}x\ +\ (3)5\ =\ (3)\,180 \qquad \text{Multiply each term of the equation by the LCD 3. Cancel out the denominators.}$$

$$3x\ +\ x\ +\ x\ +\ 15 = 540 \qquad \text{Multiply each term.}$$

$$5x\ +\ 15 = 540 \qquad \text{Combine similar terms. Subtract 15.}$$

$$5x = 525 \qquad \text{Divide by 5.}$$

$$x = 105° \qquad \text{The measure of angle } \angle A$$

$\angle B:\ \frac{1}{3}x = \frac{1}{3}(105) = 35°$ $\qquad$ $\angle C:\ \ \frac{1}{3}x + 5 = \frac{1}{3}(105) + 5 = 35 + 5 = 40°$

Answers $\angle A = 105°,\ \angle B = 35°$ and $\angle C = 40°$

(1.3)

GEOMETRY – PERIMETER OF A RECTANGLE

For any two-dimensional geometric figure, the **perimeter** is the distance around the outer edge of the figure.

The perimeter of a rectangle is equal to two lengths plus two widths:

Perimeter of a Rectangle **P = 2L + 2W**

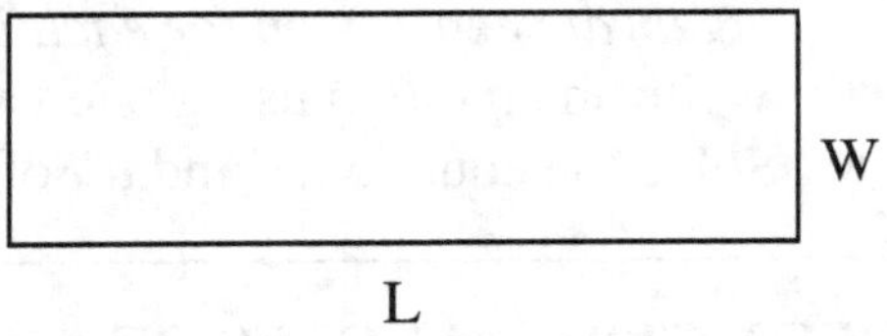

EXAMPLE 2

The length of a rectangle is 4 feet less than 3 times the width. The perimeter is 88 feet. Find the length and width.

Table	(Set-up)	(Explanation)
width	x	The length is compared to the width. Identify the width as *x*.
length	3x – 4	"4 feet less than 3 times the width"
Perimeter	88 feet	(Given in the problem)
Formula	P = 2L + 2W	
Equation	88 = 2(3x – 4) + 2(x)	
	or 2(3x – 4) + 2(x) = 88	Simplify the parentheses.
	6x – 8 + 2x = 88	(Distributive Property)
	8x – 8 = 88	Combine similar terms. Add 8.
	8x = 96	Divide by 8.
	x = 12	Simplify the answer.
Answers	x = 12 feet	Width of the rectangle
	3x – 4 = 3(12) – 4 = 32 feet	Length of the rectangle

CONSECUTIVE INTEGERS

An **integer** is a positive or negative whole number, or zero. Consecutive integers are "back-to-back," such as {4, 5, 6} or {21, 22, 23} or {58, 59, 60}. When setting up a table for consecutive integers, always denote the first number as *x*.

Helpful Hints – The set-up for three types of consecutive integer word problems:

3 Consecutive integers		**3 Consecutive even integers**		**3 Consecutive odd integers**	
1st	x	1st	x	1st	x
2nd	x + 1	2nd	x + 2	2nd	x + 2
3rd	x + 2	3rd	x + 4	3rd	x + 4

(Odd integers are 2 integers apart.)

Read the examples that follow.

(1.3)

EXAMPLE 3   The sum of three consecutive integers is 102. Find the integers.

�֍ *__Helpful Hint__* – Consecutive integers $\{\ldots, -3, -2, -1, 0, 1, 2, 3, 4, 5, \ldots\}$ are 1 unit apart.
To set up consecutive integers, add 1 to the previous number.

Table	(Set-up)	(Explanation)
1st integer	x	Always denote the first integer as x.
2nd integer	x + 1	Add 1 to the previous number.
3rd integer	x + 2	Add 1 to the previous number.
Formula	1st + 2nd + 3rd = 102	
Equation	x + x + 1 + x + 2 = 102	
	3x + 3 = 102	Combine similar terms. Subtract 3.
	3x = 99	Divide by 3.
Answers	x = 33	First integer
	x + 1 = 33 + 1 = 34	Second integer
	x + 2 = 33 + 2 = 35	Third integer

CONSECUTIVE EVEN INTEGERS

EXAMPLE 4

The sides of a triangle are given as three consecutive even integers.
The perimeter is 120 feet. Find the length of each side.

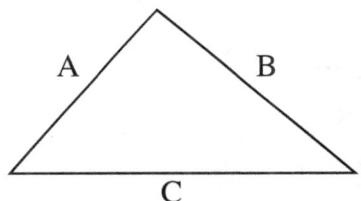

�֍ *__Helpful Hint__* – Consecutive even integers $\{\ldots, -6, -4, -2, 0, 2, 4, 6, 8, 10, \ldots\}$ are 2 units
apart. To set up consecutive even integers, add 2 to the previous number.

Table	(Set-up)	(Explanation)
side A, 1st even integer	x	Always denote the first integer as x.
side B, 2^{nc} even integer	x + 2	Add 2 to the previous number.
side C, 3rd even integer	x + 4	Add 2 to the previous number.
Perimeter	120 feet	Given in the problem
Formula	The perimeter of a triangle is equal to the sum of the three sides:	

$$P = A + B + C$$

Equation   120 = x + x + 2 + x + 4

or   x + x + 2 + x + 4 = 120

11

(1.3)

Equation	$x + x + 2 + x + 4 = 120$	Combine the similar terms.
	$3x + 6 = 120$	Subtract 6.
	$3x = 114$	Divide by 3.
Answers	$x = 38$ ft.	Side A
	$x + 2 = 38 + 2 = 40$ ft.	Side B
	$x + 4 = 38 + 4 = 42$ ft.	Side C

CONSECUTIVE ODD INTEGERS

EXAMPLE 5   If the largest of three consecutive odd integers is decreased by the smallest, the result is equal to one-fifth of the middle integer minus 3. Find the integers.

❈ *Helpful Hint* – Consecutive **odd** integers $\{\ldots, -5, -3, -1, 1, 3, 5, 7, 9, 11, \ldots\}$ are **2 units apart**. To set up consecutive odd integers, add 2 to the previous number.

Table	(Set-up)	(Explanation)
(smallest) 1st odd integer	x	Always denote the first integer as x.
(middle) 2nd odd integer	$x + 2$	Add 2 to the previous number.
(largest) 3rd odd integer	$x + 4$	Add 2 to the previous number.

Formula   "Largest decreased by smallest is equal to one-fifth of the middle integer minus 3"

$$\text{Lg} \quad - \quad \text{Sm} \quad = \quad \frac{1}{5} \text{ of Mid} \quad - \quad 3$$

Equation	$x + 4 - x = \frac{1}{5}(x + 2) - 3$	Combine similar terms on the left side.
	$4 = \frac{1}{5}(x + 2) - 3$	Simplify the parentheses.
	$4 = \frac{1}{5}x + \frac{2}{5} - 3$	Multiply each term by the LCD 5.
	$(5)4 = (5)\frac{1}{5}x + (5)\frac{2}{5} - (5)3$	Simplify each term.
	$20 = x + 2 - 15$	Combine similar terms.
	$20 = x - 13$	Add 13.
Answers	$x = 33$	1st odd integer
	$x + 2 = 33 + 2 = 35$	2nd odd integer
	$x + 4 = 33 + 4 = 37$	3rd odd integer

Check: Lg $-$ Sm $= \frac{1}{5}$ of Mid $- 3$ $\rightarrow$ $37 - 33 = \frac{1}{5}(35) - 3$ $\rightarrow$ $4 = 7 - 3$ is true.

SIMPLE INTEREST (INVESTMENTS)

If a one-time deposit of money is invested in a financial account, the interest earned at the end of one year is determined by: Interest for 1 year = (interest rate)(amount invested)

EXAMPLE 6   Bill and Janna Brown invested a total of $20,000 in two accounts. One account pays 7% per year and the other account pays 9% per year. The total interest earned in the first year was $1,660. How much money was invested in each account?

�֎ *Helpful Hints* – (1) Observe that amount invested in the 7% account was <u>not</u> compared to the amount invested in the 9% account. In general, if the two items are not compared, then identify one of the items as x. Describe the other item as the "Total amount – x."   (2) When you set up the interest rates, write each percent in decimal form.

Table	(Set-up)	(Explanation)
Amount invested, 7% Account	x	Denote the first amount invested as x.
Amount invested, 9% Account	20,000 – x	Denote the second amount as "Total – x."
Interest, 7% Account	0.07(x)	Interest earned = 1st Rate (1st Amount)
Interest, 9% Account	0.09(20,000 – x)	Interest earned = 2nd Rate (2nd Amount)
Total Interest, first year	$1,660	(Given in the problem)

Formula   "Interest on 1st Acct   +   Interest on 2nd Acct   = Total Interest Earned"

$\qquad$ 1st Rate (1st Amount) + 2nd Rate (2nd Amount) = Total Interest

Equation	$0.07(x) + 0.09(20,000 - x) = 1,660$	
	$0.07x + 1,800 - 0.09x = 1,660$	Simplify the parentheses.
	$-0.02x + 1,800 = 1,660$	Combine similar terms. Subtract 1800.
	$-0.02x = -140$	Divide by -0.02 on both sides.
Answers	$x = \$7,000$	Amount invested in the 7% Account
	$20,000 - x = 20,000 - 7,000 = \$13,000$	Amount invested in the 9% Account

Check:

Interest, 7% Account	0.07 ($7,000) = $490
Interest, 9% Account	0.09($13,000) = $1,170
Total Interest earned	$490 + $1,170 = $1,660

(1.3)

MIXTURE PROBLEM

EXAMPLE 7   A chemical storage room has one container of a 15% acid solution and another container of a 55% acid solution. How many liters of each solution must be mixed to make 12 liters of a 25% acid solution?

�֎ _Helpful Hints_ - (1) Observe that amount of the 15% solution was not compared to the amount of the 55% solution. Recall (from Example 6) that the amount of one solution will be identified as x and the amount of the other solution will be described as the "Total amount – x." (2) When you set up the equation, write each percent in decimal form.

Table	(Set-up)	(Explanation)
Amount of 15% acid solution	x	Denote the first amount invested as x.
Amount of 55% acid solution	12 – x	Denote the second amount as "Total – x."
Acid in the 15% solution	0.15(x)	Acid = Percentage (1st Amount)
Acid in the 55% solution	0.55(12 – x)	Acid = Percentage (2nd Amount)
Acid in the 25% mixture	0.25(12)	Acid = Percentage (Total Amount)

Formula   "Acid in the 15% solution + Acid in the 55% solution = Acid in the 25% mixture"

Equation $$0.15(x) + 0.55(12 - x) = 0.25(12)$$

$0.15x + 6.6 - 0.55x = 3$	Simplify the parentheses.
$-0.4x + 6.6 = 3$	Combine similar terms. Subtract 6.6.
$-0.4x = -3.6$	Divide by -0.04 on both sides.

Answers $x = 9$ liters   Amount of 15% acid solution

$12 - x = 12 - 9 = 3$ liters   Amount of 55% acid solution

MOTION PROBLEMS

Applications which involve two vehicles moving in opposite directions, or, a traveler making a round trip are called motion problems. These problems use the **distance formula**:

Distance traveled = (Rate or speed of the vehicle)(Time traveled)

or

$$D = rt$$

Therefore, if a plane flies at a rate of 200 miles per hour and travels for 3 hours, the distance traveled by the plane would be: (200 mph)(3 hrs) = 600 miles

14

EXAMPLE 8

A car and a minivan take off from an intersection on a state highway and drive in opposite directions. The minivan travels an average of 7 mph slower than the car. After 2 hours, the vehicles are 234 miles apart. Find the speed of each vehicle.

✠ *Helpful Hint* – This example is an **opposite direction** problem. Observe that the distance traveled by the car plus the distance traveled by the minivan equals the total distance between them:

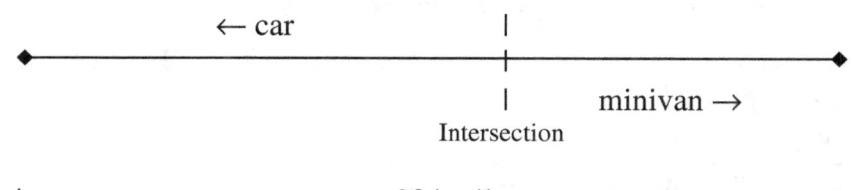

Distance for car + Distance for minivan = Total distance apart

Table	(Set-up)	(Explanation)
Rate of the car	x	Identify the rate of the car as *x*.
Rate of the minivan	x – 7	"7 mph slower than the car"
Distance traveled by the car	(x)(2 hrs)	Distance = (rate)(time)
Distance traveled by the minivan	(x – 7)(2 hrs)	Distance = (rate)(time)
Total distance apart	234 miles	(Given in the problem)

Formula   Distance for car + Distance for minivan = Total distance apart

Equation $(x)(2) + (x - 7)(2) = 234$

$$2x + 2x - 14 = 234$$ Simplify the parentheses.

$$4x - 14 = 234$$ Combine similar terms. Add 14.

$$4x = 248$$ Divide by 4.

Answers $x = 62$ mph   Speed (rate) of the car

$x - 7 = 62 - 7 = 55$ mph   Speed (rate) of the minivan

Check:

The distance traveled by the car   (x)(2 hrs) = (62 mph)(2 hrs) = 124 miles

The distance traveled by the minivan   (x – 7)(2 hrs) = (55 mph)(2 hrs) = 110 miles

Total distance apart   124 + 110 = 234 miles

(1.3)

EXAMPLE 9

Diane jogged from her home to a community park at a speed of 4 miles per hour. Then she decided to walk home from the park, along the same route, at 2.5 miles per hour. It took her 18 minutes longer to walk home than to jog to the park. (a) How long did it take Diane to jog to the park? (b) How long did it take her to walk home? (c) How far is the community park from her home?

✠ *__Helpful Hints__* – (1) This example is a **round trip** problem. Observe that the distance to the park is the same as the distance traveled back to Diane's home. In general, for a round trip problem:

$$\text{Distance Going To} = \text{Distance Coming Back}$$

(2) Also observe that the speeds or travel rates are given as miles per *hour*. The "18 minutes longer" for the return trip must be changed to hours:

$$18 \text{ minutes} = \left(\frac{18 \text{ min}}{1}\right)\left(\frac{1 \text{ hr}}{60 \text{ min}}\right) = \left(\frac{18}{60}\right)\text{hr} = \frac{3}{10}\text{hr} = 0.3\text{ hr}$$

Table	(Set-up)	(Explanation)
Time going to the park	x hrs	Identify the time going to the park as *x*.
Time coming back home	(x + 0.3) hrs	"18 minutes longer" stated in hours
Distance going to the park	(4)(x)	Distance = (rate)(time)
Distance coming back home	(2.5)(x + 0.3)	Distance = (rate)(time)

Formula   Distance Going To = Distance Coming Back

Equation		
$(4)(x) = (2.5)(x + 0.3)$		
$4x = 2.5x + 0.75$	Simplify the parentheses.	
$1.5x = 0.75$	Subtract 2.5x on both sides of the equation.	
$x = \dfrac{0.75}{1.5}$	Divide by 1.5 on both sides.	
$x = 0.5 \text{ hr}$	Time going to the park (in hours)	
$x + 0.3 = 0.5 + 0.3 = 0.8 \text{ hr.}$	Time coming back home (in hours)	

In minutes:

x = 0.5 hour = ½ hour = 30 minutes	Time going to the park
30 min + 18 min longer = 48 minutes	Time coming back home

Check: (0.8 hr)(60 minutes per hour) = 48 minutes

Answers   (a) It took Diane one-half hour or 30 minutes to jog from her home to the park.
         (b) It took her 48 minutes to walk home from the park.
         (c) The distance from her home to the park is:
             Distance = (rate)(time) = (4 mph)(0.5 hr) = 2 miles

16

1.4   LINEAR INEQUALITIES IN ONE VARIABLE

GRAPHING INEQUALITIES ON THE NUMBER LINE

An inequality is a mathematical expression which contains an inequality symbol:

| > Greater than | $7 > 4$ | ≥ Greater than or equal | $7 \geq 4$ | $2 + 2 \geq 4$ |
| < Less than | $3 < 5$ | ≤ Less than or equal | $3 \leq 5$ | $2 + 3 \leq 5$ |

The real **number line** below has an arrow at each end since a number line is infinite. It also has equally spaced scale marks.

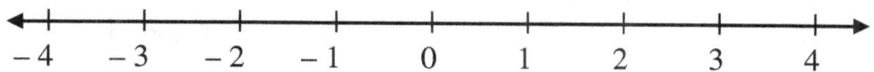

The arrow at the left end indicates **negative infinity**.   " $-\infty$ "
The arrow at the right end indicates **positive infinity**.   " ∞ "

An inequality which uses a variable describes a set of real numbers called an **interval**.

> **EXAMPLE 1**   (a) The inequality $x > 1$ is the set of all real numbers which are greater than 1. The inequality $x > 1$ also indicates the interval "all real numbers above 1 to infinity, *not including* the endpoint 1."
>
> (b) The inequality $x \geq 1$ is the set of all real numbers which are greater than 1 or equal to 1. The inequality $x \geq 1$ also indicates the interval "all real numbers from 1 to infinity, *including* the endpoint 1."

Interval Notation is used to describe the shaded part of a number line graph from left to right. In the following table, the variable *n* represents the endpoint of the inequality interval. For interval notation, a comma indicates "up to."

Inequality Notation	Graphing Symbols		Interval Notation
$x > n$	⟵⟶ or o⟶		(n, ∞)
$x \geq n$	⊢⟶ or ●⟶		$[n, \infty)$
$x < n$	⟵) or ⟵o		$(-\infty, n)$
$x \leq n$	⟵⊣ or ⟵●		$(-\infty, n]$

EXAMPLE 2   The following illustrations show the graph and the interval notation for each inequality.

Inequality Notation	Graph	Interval Notation
(a) $x > 1$	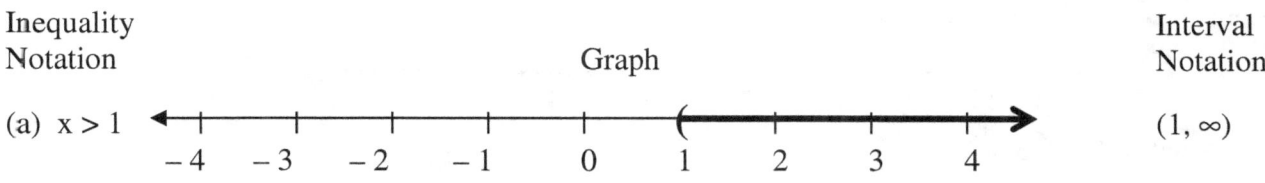	$(1, \infty)$

(1.4)

Inequality Notation	Graph	Interval Notation

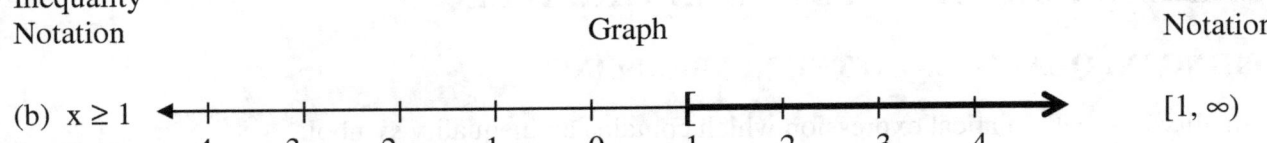

(b) x ≥ 1   [1, ∞)

(c) x < 1   (− ∞, 1)

(d) x ≤ 1   (− ∞, 1]

LINEAR INEQUALITIES

A linear inequality has the variable in the first degree. (i.e. There are no exponents.)

> **SPECIAL INEQUALITY RULE:** If you divide or multiply both sides of an inequality by a negative number, change the inequality symbol.

EXAMPLE 3   (Numerical Illustrations)

For Most Operations	Special Inequality Rule
(a) Given: − 2 < 6	(b) Given: − 2 < 6
$\quad\quad\quad\quad\quad\quad\dfrac{-2}{2} < \dfrac{6}{2}$	$\quad\quad\quad\quad\quad\quad\dfrac{-2}{-2} < \dfrac{6}{-2}$
Divide By 2:	Divide By − 2:
Result: − 1 < 3	Result: 1 > − 3
(The inequality symbol stays the same.)	*(Change the inequality symbol.)*

SOLVING LINEAR INEQUALITIES

EXAMPLE 4   Solve: 7x − 9 > 5x − 3

$$\quad\quad\quad\quad\quad − 5x \quad\quad\quad − 5x \quad\quad\quad\text{Subtract 5x from both sides of the inequality.}$$

$$\quad\quad\quad\quad\quad 2x − 9 > − 3$$

$$\quad\quad\quad\quad\quad\quad\quad + 9 \quad\quad + 9 \quad\quad\quad\text{Add 9 to each side of the inequality.}$$

$$\quad\quad\quad\quad\quad\quad\quad 2x > 6$$

$$\quad\quad\quad\quad\quad\quad\quad \frac{2x}{2} > \frac{6}{2} \quad\quad\quad\quad\text{Divide both sides by 2.}$$

Solution: x > 3

Graph:

Interval Notation: (3, ∞)

(1.4)

EXAMPLE 5   Solve: $4(x - 1) - 7x \geq 1 - 2(x - 5) + 5x$

$$4x - 4 - 7x \geq 1 - 2x + 10 + 5x$$ Simplify the parentheses using the Distributive Property.

$$-3x - 4 \geq 3x + 11$$ Combine the similar terms on each side.
$$-3x \qquad -3x$$ Subtract 3x from both sides of the inequality.

$$-6x - 4 \geq 11$$
$$+4 \qquad +4$$ Add 4 to each side of the inequality.

$$\frac{-6x}{-6} \geq \frac{15}{-6}$$ Divide both sides by -6.

Solution: $x \leq -\dfrac{5}{2}$ Change the inequality symbol (Special Inequality Rule).

Graph: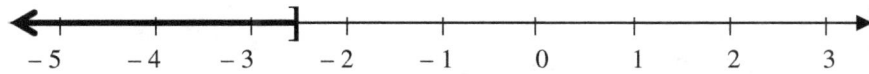

$$-5 \quad -4 \quad -3 \quad -2 \quad -1 \quad 0 \quad 1 \quad 2 \quad 3$$

Interval Notation: $\left(-\infty, -\dfrac{5}{2}\right]$

CONTINUED INEQUALITIES   (The "Between" Statement)

The inequality $-1 \leq x \leq 2$ means "all numbers **between** -1 and 2, including -1 and 2."

Graph: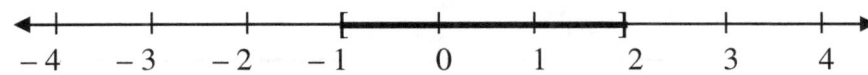

$$-4 \quad -3 \quad -2 \quad -1 \quad 0 \quad 1 \quad 2 \quad 3 \quad 4$$

Interval Notation: $[-1, 2]$

EXAMPLE 6   Solve for x: $-18 < 5x - 4 \leq 2$

✖ *__Helpful Hint__* – Observe that the continued inequality has three parts: the left, the middle, and the right. Whatever operation is used on one part of the inequality must be applied to all three parts.

$$-18 < 5x - 4 \leq 2$$
$$+4 \qquad\quad +4 \quad\; +4$$ Add 4 to each part of the inequality.

$$-14 < 5x \leq 6$$

$$\frac{-14}{5} < \frac{5x}{5} \leq \frac{6}{5}$$ Divide each part by 5.

Solution: $\dfrac{-14}{5} < x \leq \dfrac{6}{5}$ i.e. $-2\dfrac{4}{5} < x \leq 1\dfrac{1}{5}$

Graph: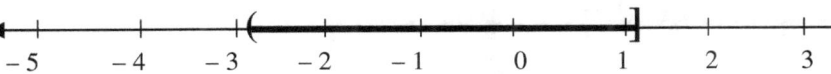

$$-5 \quad -4 \quad -3 \quad -2 \quad -1 \quad 0 \quad 1 \quad 2 \quad 3$$

Interval Notation: $\left(-\dfrac{14}{5}, \dfrac{6}{5}\right]$

1.5   COMPOUND INEQUALITIES

A compound inequality consists of two or more inequalities (subsets) joined by the word *And* or *Or*.

Intersection (The "And" Statement)

For a compound inequality, "And" means the solution is the set of numbers which make *both* subsets true. The answer is the **intersection** (i.e. the overlap) of both subsets. The symbol for intersection is "∩". Therefore, the solution set may be written as a continued inequality:

$$\{x \geq -1\} \cap \{x \leq 2\} \rightarrow \{x \geq -1 \text{ and } x \leq 2\} \rightarrow \{-1 \leq x \text{ and } x \leq 2\} \rightarrow \{-1 \leq x \leq 2\}$$

In other words, the *And* inequality statement may be rewritten as a *Between* inequality statement.

For the *And* compound inequality, solve each inequality separately. Then write the solution subsets as a continued inequality.

EXAMPLE 1   Solve: $2x + 5 > 9$ and $3x + 5 < 20$

$2x + 5 > 9$	and	$3x + 5 < 20$	Solve each inequality separately.
$2x > 4$		$3x < 15$	
$x > 2$	and	$x < 5$	Write the solution subsets.
$2 < x$	and	$x < 5$	Rewrite the left solution subset.

Solution: $2 < x < 5$ Write the solution subsets as a continued inequality.

Graph:

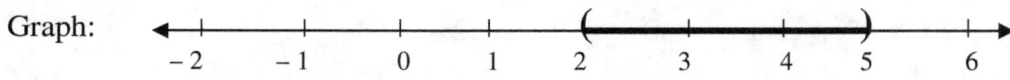

Interval Notation: (2, 5)

Union (The "Or" Statement)

For a compound inequality, "Or" means the solution is the set of numbers which make *either* subset true. Solve each inequality separately. The answer is **union** of the solution subsets. This means graph *both* subsets on the *same number line*.

EXAMPLE 2   Solve: $5x - 17 \leq -2$ or $5x - 17 \geq 3$

$5x - 17 \leq -2$	or	$5x - 17 \geq 3$	Solve each inequality separately.
$5x \leq 15$		$5x \geq 20$	
$x \leq 3$		$x \geq 4$	

Solution: $x \leq 3$ or $x \geq 4$

Graph:

In interval notation, the symbol for *Or* is the union symbol "∪".

Interval Notation: $(-\infty, 3] \cup [4, \infty)$

1.6   REVIEW OF SQUARE ROOTS

DEFINITIONS FOR REAL SQUARE ROOTS

The expression "square root of x" is denoted as $\sqrt[2]{x}$ or $\sqrt{x}$. The number "2" at the upper left is called the **index** and the expression under the root symbol, x, is called the **radicand**. A square root is usually written without the index. To get a real number answer for a square root, the radicand must be positive or zero.

EXAMPLE 1   (a) $\sqrt{25} = 5$, a real number.   (b) $-\sqrt{25} = -5$, a real number.

DEFINITIONS FOR IMAGINARY NUMBERS

The square root of a negative real number is called an **imaginary number**. Thus $\sqrt{-25}$ is an imaginary number.

The following are two important Rules for Imaginary Numbers.
- The expression $\sqrt{-1}$ must be changed to the **imaginary unit** i: $\sqrt{-1} = i$
- Let "a" be any positive real number. The square root of a negative number must rewritten as an imaginary number with a positive radicand: $\sqrt{-a} = i\sqrt{a}$

EXAMPLE 2   (a) $\sqrt{-25} = \sqrt{-1}\sqrt{25} = (i)(5) = 5i$ (b) $\sqrt{-4} = \sqrt{-1}\sqrt{4} = (i)(2) = 2i$

(c) *SHORTCUT*: $\sqrt{-9} = 3i$ since $\sqrt{9} = 3$ and $\sqrt{-1} = i$

SIMPLIFYING SQUARE ROOTS

EXAMPLE 3 $\sqrt{-81} + \sqrt{-144} - \sqrt{-49} = 9i + 12i - 7i = 14i$

EXAMPLE 4 $\dfrac{\sqrt{-400}}{\sqrt{16}} = \dfrac{20i}{4} = 5i$

EXAMPLE 5 $\dfrac{\sqrt{144} + \sqrt{-100}}{\sqrt{169} - \sqrt{121}} = \dfrac{12 + 10i}{13 - 11} = \dfrac{12 + 10i}{2} = \dfrac{12}{2} + \dfrac{10i}{2} = 6 + 5i$

EXAMPLE 6   For the triangle at the right, a = 2.5 inches,

$b = \dfrac{5\sqrt{3}}{2}$ inches, and c = 5 inches.

Find the perimeter (to the nearest hundredth) using the formula P = a + b + c.

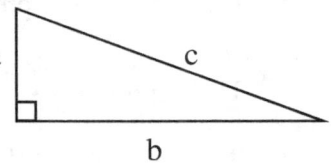

Answer: Using a calculator, Perimeter P = $2.5 + \dfrac{5\sqrt{3}}{2} + 5 = 2.5 + 4.330127019 + 5$

$= 11.830127019 \approx 11.83$ inches

21

(1.6)

EXAMPLE 7   For the triangle at the right, a = 2.5 inches,

b = $\dfrac{5\sqrt{3}}{2}$ inches, and c = 5 inches.

Find the area (to the nearest hundredth) using

the formula A = $\dfrac{1}{2}$ bh.

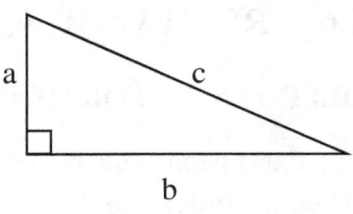

Answer: Using a calculator, Area A = $\left(\dfrac{1}{2}\right)\left(\dfrac{5\sqrt{3}}{2}\right)(2.5)$ = $\dfrac{\left(5\sqrt{3}\right)(2.5)}{4}$ = $\dfrac{21.65063509}{4}$

$= 5.412658774 \approx 5.41 \text{ in}^2$

EXAMPLE 8   Graph the inequality $x \geq -\sqrt{2}$ on a standard number line. Round the decimal to the nearest tenth.

Answer: Using a calculator, $-\sqrt{2} = -1.414213562\ldots \approx -1.4$ (rounded). Rewrite the inequality as: $x \geq -1.4$

Graph:

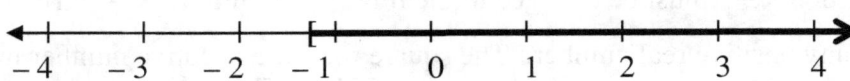

EXAMPLE 9   Graph the compound inequality $x < \left(1 - 2\sqrt{2}\right)$ or $x > \left(1 + 2\sqrt{2}\right)$ on a standard number line. Round the decimals to the nearest tenth.

Answer: The expression $1 - 2\sqrt{2} = -1.828427125\ldots \approx -1.8$ (rounded).
The expression $1 + 2\sqrt{2} = 3.828427125\ldots \approx 3.8$ (rounded). Rewrite the compound inequality as: $x < -1.8$ or $x > 3.8$

Graph:

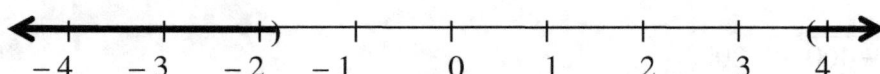

EXAMPLE 10   Graph the inequality $\dfrac{6 - \sqrt{61}}{5} \leq x \leq \dfrac{6 + \sqrt{61}}{5}$ on a standard number line. Round the decimals to the nearest tenth.

Answer: The expression $\dfrac{6 - \sqrt{61}}{5} = -0.362049935\ldots \approx -0.4$ (rounded).

The expression $\dfrac{6 + \sqrt{61}}{5} = 2.762049935\ldots \approx 2.8$ (rounded). Rewrite the compound inequality as: $-0.4 \leq x \leq 2.8$

Graph:

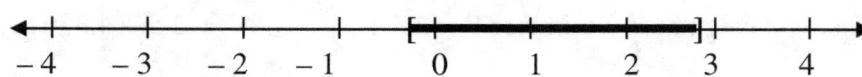

Chapter 2   Rectangular Coordinates and Functions

2.1   RECTANGULAR COORDINATES AND BASIC GRAPHS

An **ordered pair** consists of two numbers written in parentheses and separated by a comma. The order of the numbers effects the meaning of the ordered pair so that (2, 3) is not the same as (3, 2).

The grid at the lower right is called a **rectangular coordinate system.** The bold horizontal number line is the **x-axis**. The bold vertical number line is the **y-axis**. The point where both axes cross is called the **origin**. Each point in the system is named by an ordered pair (x, y). The first number is the **x-coordinate** and the second number is the **y-coordinate**. To graph a point, start at the origin and count x units across, y units vertically and place a point (small dot) when you finish the second number.

EXAMPLE 1   Graph each point and label it:

A. (4, 4)   B. (3, – 3)   C. (– 4, 2)

D. (– 1, – 5)   E. (2, 0)   F. (0, – 2)

G. (0, 0)

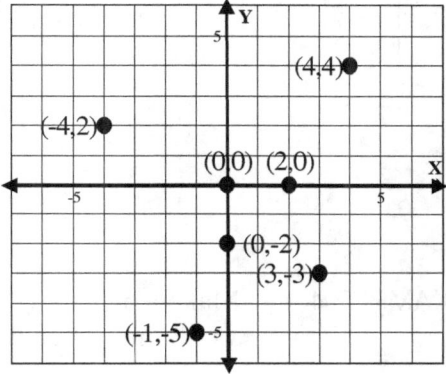

GRAPHS OF BASIC EQUATIONS IN TWO VARIABLES

The **linear equation** has the format **y = ax + b** where **a** and **b** are real numbers. The linear equation has no exponents. The graph of a linear equation is a straight line (no corners, no curves).

EXAMPLE 2   Make a 5-point table of ordered pairs using the integers of the interval [-2, 2] in the x column. Then draw the graph for y = 2x – 1.

x	y = 2x – 1
– 2	– 5
– 1	– 3
0	– 1
1	1
2	3

(Workspace)

$y = 2(– 2) – 1 = – 5$

$y = 2(– 1) – 1 = – 3$

$y = 2(0) – 1 = – 1$

$y = 2(1) – 1 = 1$

$y = 2(2) – 1 = 3$

Graph:

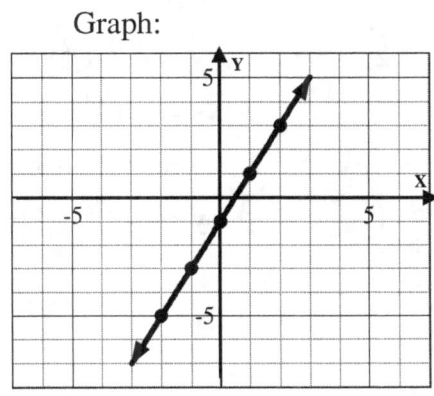

23

(2.1)

The remaining equations of this section are *nonlinear*. The graphs are *not* straight lines.

The **square equation**: The graph of $y = x^2$ is a **parabola** and the **vertex** or lowest point is the origin $(0, 0)$. Also, place an arrow at the left and right ends of the graph.

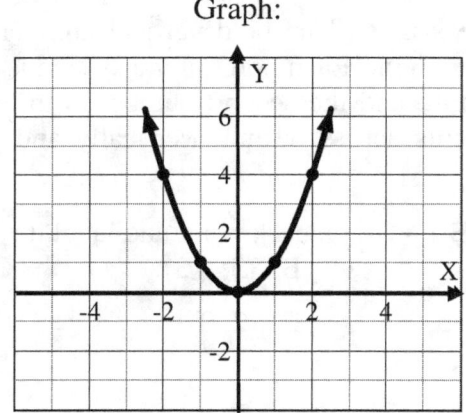

EXAMPLE 3   Make a 5-point table of ordered pairs using the integers of the interval [-2, 2] in the x column. Then draw the graph for $y = x^2$.

x	$y = x^2$	(Workspace)
-2	4	$y = (-2)^2 = 4$
-1	1	$y = (-1)^2 = 1$
0	0	$y = (0)^2 = 0$
1	1	$y = (1)^2 = 1$
2	4	$y = (2)^2 = 4$

Graph:

EXAMPLE 4   Make a 5-point table of ordered pairs using the integers of the interval [-2, 2] in the x column. Then draw the graph for $y = -x^2 + 3$.

The equation $y = -x^2 + 3$ has a **leading coefficient** of -1 (the number in front of the x^2 term).

✳ *__Helpful Hint__* – For any equation in two variables x and y, let y equal a mathematical expression in terms of x. If the *leading coefficient* of the expression is *negative*, the graph will be inverted or upside-down. The inverted graph is called a **reflection**.

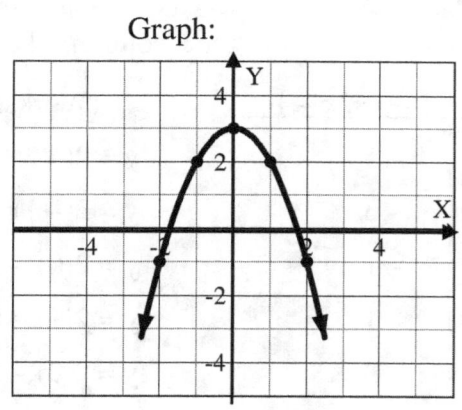

x	$y = -x^2 + 3$	(Workspace)
-2	-1	$y = -(-2)^2 + 3 = -4 + 3 = -1$
-1	2	$y = -(-1)^2 + 3 = -1 + 3 = 2$
0	3	$y = -(0)^2 + 3 = -0 + 3 = 3$
1	2	$y = -(1)^2 + 3 = -1 + 3 = 2$
2	-1	$y = -(2)^2 + 3 = -4 + 3 = -1$

Graph:

(2.1)

The **absolute value equation**: The graph of $y = |x|$ is **2 rays**.
The **vertex** is the corner point located at the origin $(0, 0)$.
Place an arrow at the left and right ends of the graph.

EXAMPLE 5   Make a 5-point table of ordered pairs using the
integers of the interval [-2, 2] in the x column. Then draw the graph
for $y = |x|$.

x	$y = |x|$
– 2	2
– 1	1
0	0
1	1
2	2

(Workspace)

$y = |-2| = 2$

$y = |-1| = 1$

$y = |0| = 0$

$y = |1| = 1$

$y = |2| = 2$

Graph:

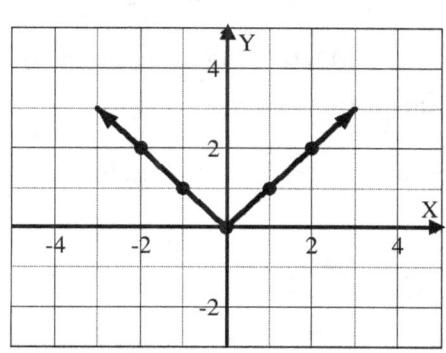

The **cubic equation**: The graph of $y = x^3$ has the shape of curve at the right.
The **center** point is the origin $(0, 0)$. Place an arrow at both ends of the graph.

EXAMPLE 6   Make a 5-point table of ordered pairs using the integers of
the interval [-2, 2] in the x column. Then draw the graph for $y = x^3$.

x	$y = x^3$
– 2	– 8
– 1	– 1
0	0
1	1
2	8

(Workspace)

$y = (-2)^3 = -8$

$y = (-1)^3 = -1$

$y = (0)^3 = 0$

$y = (1)^3 = 1$

$y = (2)^3 = 8$

Graph:

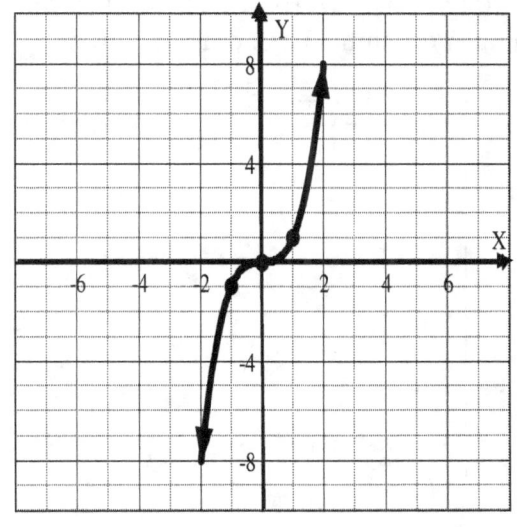

(2.1)

The **square root equation**: The graph of $\mathbf{y} = \sqrt{x}$ is the top half of a right parabola.

Remember that x-coordinates and y-coordinates must be real numbers.
The **starting point** of the graph is the origin (0, 0).

✠ *__Helpful Hint__* – To find the x-coordinate of the starting point for a square root equation, set the radicand greater than or equal to zero and solve for *x*.

For the equation $\mathbf{y} = \sqrt{x}$, we set the radicand $x \geq 0$. Thus the x-coordinate of the starting point is zero.

EXAMPLE 7   Make a 5-point table using consecutive integers starting with 0 in the *x* column. Then draw the graph for $y = \sqrt{x}$.

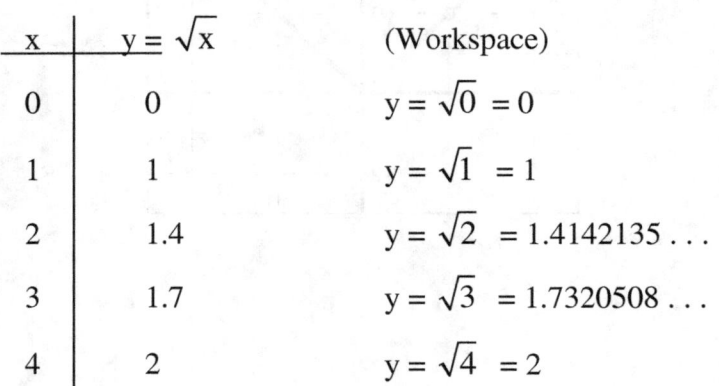

x	$y = \sqrt{x}$	(Workspace)
0	0	$y = \sqrt{0} = 0$
1	1	$y = \sqrt{1} = 1$
2	1.4	$y = \sqrt{2} = 1.4142135\ldots$
3	1.7	$y = \sqrt{3} = 1.7320508\ldots$
4	2	$y = \sqrt{4} = 2$

Graph:

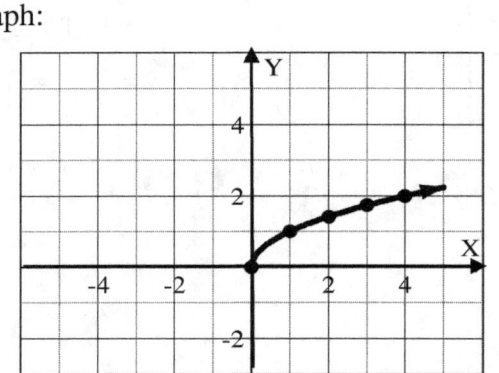

✠ *__Helpful Hint__* – Observe that the graph of the square root equation has a *point* at one end and *an arrow* at the other end.

EXAMPLE 8   For the equation $\mathbf{y} = -\sqrt{x}$, we set the radicand $x \geq 0$. Thus the x-coordinate of the starting point is zero. Use this result to make a 5-point table of ordered pairs with consecutive integers in the *x* column. Then draw the graph.

x	$y = -\sqrt{x}$	(Workspace)
0	0	$y = -\sqrt{0} = 0$
1	-1	$y = -\sqrt{1} = -1$
2	-1.4	$y = -\sqrt{2} = -1.4142135\ldots$
3	-1.7	$y = -\sqrt{3} = -1.7320508\ldots$
4	-2	$y = -\sqrt{4} = -2$

Graph:

(2.1)

EXAMPLE 9   Let $y = \sqrt{x+2}$. To find the x-coordinate for the starting point, set the radicand $x + 2 \geq 0$ and solve for x:

$$x + 2 \geq 0$$
$$x \geq -2 \qquad \text{The x-coordinate for the starting point is } -2.$$

Make a 5-point table of ordered pairs with consecutive integers in the x column. Then draw the graph.

x	y	(Workspace)
-2	0	$y = \sqrt{-2+2} = \sqrt{0} = 0$
-1	1	$y = \sqrt{-1+2} = \sqrt{1} = 1$
0	1.4	$y = \sqrt{0+2} = \sqrt{2} = 1.4142135\ldots$
1	1.7	$y = \sqrt{1+2} = \sqrt{3} = 1.7320508\ldots$
2	2	$y = \sqrt{2+2} = \sqrt{4} = 2$

Graph:

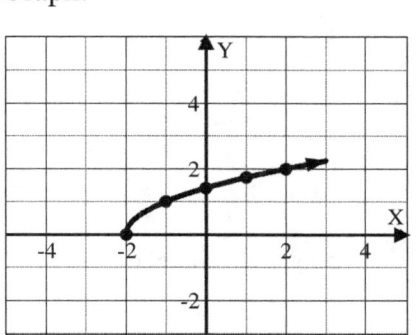

The **rational equation**: The graph of $y = \dfrac{1}{x}$ consists of two curves called a **hyperbola**.

Remember that the denominator of a fraction cannot be zero. Therefore this equation is undefined when x is zero, which is denoted as $x \neq 0$. When we graph the hyperbola, one curve is to the left of the y-axis and one curve is to the right of the y-axis. *The curves are <u>not</u> connected.*

EXAMPLE 10   Make a table of ordered pairs using numbers from the interval $[-3, 3]$ in the x column. Then draw the graph for $y = \dfrac{1}{x}$.

x	$y = \dfrac{1}{x}$	(Workspace)
-3	$-\dfrac{1}{3}$	$y = \dfrac{1}{-3} = -\dfrac{1}{3}$
-2	$-\dfrac{1}{2}$	$y = \dfrac{1}{-2} = -\dfrac{1}{2}$
-1	-1	$y = \dfrac{1}{-1} = -1$
-0.5	-2	$y = \dfrac{1}{-0.5} = -2$
0	(no y-coordinate!)	$y = \dfrac{1}{0} = \text{undefined}$
0.5	2	$y = \dfrac{1}{0.5} = 2$
1	1	$y = \dfrac{1}{1} = 1$
2	$\dfrac{1}{2}$	$y = \dfrac{1}{2}$
3	$\dfrac{1}{3}$	$y = \dfrac{1}{3}$

Graph:

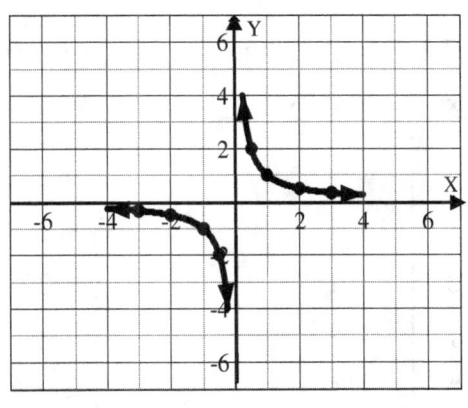

(2.1)

EXAMPLE 11   Make a table of ordered pairs using numbers from the interval $[-3, 3]$ in the x column. Then draw the graph for $y = -\dfrac{1}{x}$.

x	$y = -\dfrac{1}{x}$
-3	$\dfrac{1}{3}$
-2	$\dfrac{1}{2}$
-1	1
-0.5	2
0	(no y-coordinate!)
0.5	-2
1	-1
2	$-\dfrac{1}{2}$
3	$-\dfrac{1}{3}$

(Workspace)

$y = -\dfrac{1}{-3} = \dfrac{1}{3}$

$y = -\dfrac{1}{-2} = \dfrac{1}{2}$

$y = -\dfrac{1}{-1} = 1$

$y = -\dfrac{1}{-0.5} = 2$

$y = -\dfrac{1}{0} =$ undefined

$y = -\dfrac{1}{0.5} = -2$

$y = -\dfrac{1}{1} = -1$

$y = -\dfrac{1}{2}$

$y = -\dfrac{1}{3}$

Graph:

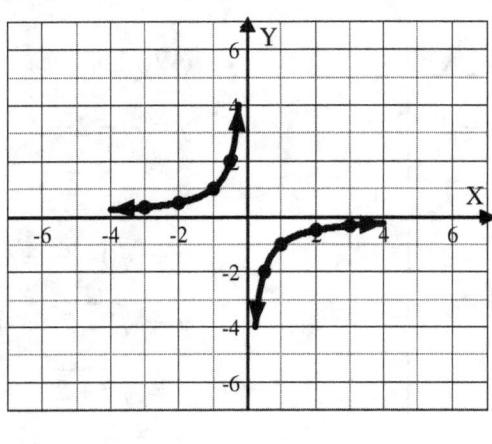

2.2   FUNCTIONS

A **relation** is: (1) a set of ordered pairs (x, y) *or* (2) an equation in two variables which generates a set of ordered pairs   *or* (3) the graph of a set of ordered pairs.

EXAMPLE 1   Show that the equation $y = 2x + 1$ is a relation by making a table of ordered pairs.

x	$y = 2x + 1$
$\vdots$	$\vdots$
-2	-3
-1	-1
0	1
1	3
2	5
$\vdots$	$\vdots$

(Workspace)

$y = 2(-2) + 1 = -3$

$y = 2(-1) + 1 = -1$

$y = 2(0) + 1 = 1$

$y = 2(1) + 1 = 3$

$y = 2(2) + 1 = 5$

Graph:

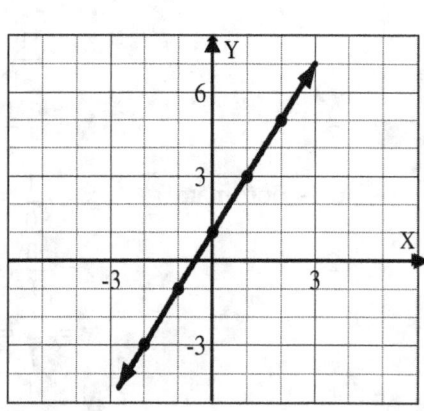

Domain: All of the x-coordinates.
Range: All of the y-coordinates.

For a relation, the **domain** is the set of all *possible x-coordinates* and the **range** is the set of all *possible y-coordinates*.

IDENTIFYING A FUNCTION

A **function** is a relation for which *each x* in the domain corresponds to *exactly one y* in the range.

Algebra II Examples: Given A Set of Ordered Pairs

EXAMPLE 2   Determine whether the relation $\{(-4, 17), (4, 17), (5, 26), (6, 37)\}$ is a function.

Workspace:

x	y
-4	17
4	17
5	26
6	37

Each x-coordinate is paired with exactly one y-coordinate.
Answer: This relation is a function.

EXAMPLE 3   Determine whether the relation $\{(-1, 0), (0, -1), (3, -2), (0, 1), (3, 2)\}$ is a function.

Workspace:

The x-coordinate -1 is paired with 0. But 0 is paired with -1 and 1. Also, 3 is paired with -2 and 2.
Answer: This relation is a not function.

College Algebra Examples: Given An Equation In Two Variables

Solve for y. Then determine whether each x-coordinate has exactly one y-coordinate.

EXAMPLE 4   Is the relation $y = -4x + 1$ a function?

Workspace:

x	y	
$\vdots$	$\vdots$	
-2	9	$-4(-2) + 1$
0	1	$-4(0) + 1$
5	-19	$-4(5) + 1$
$\vdots$	$\vdots$	

Answer: Each x-coordinate that we choose will have exactly one y-coordinate.
The equation $y = -4x + 1$ is a function.

(2.2)

EXAMPLE 5   Is the relation $y^2 = x + 1$ a function?

Solve for y. Then determine whether each x-coordinate has exactly one y-coordinate.

Workspace: $\sqrt{y^2} = \pm\sqrt{x+1}$ For a 2nd degree equation: Write the square root of both sides.

$y = \pm\sqrt{x+1}$ Use "double signs" on the right side of the equation.

x	y	
-1	0	$\pm\sqrt{-1+1}$
0	± 1	$\pm\sqrt{0+1}$

Answer: Observe that x = 0 has two y-coordinates, y = 1 and y = $-$1.
This equation $y^2 = x + 1$ is not a function.

College Algebra Example: Given the Graph of a Relation

Use the **Vertical Line Test:** *The graph of an relation represents a function if each vertical line intersects the graph in at most one point.*

✖️ *Helpful Hint* – If any vertical line intersects the graph more than once, the graph is not a function.

EXAMPLE 6   Use the Vertical Line Test to determine whether each graph below represents a function.

Graph A   Graph B

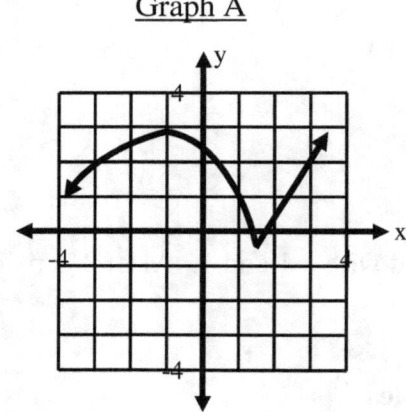   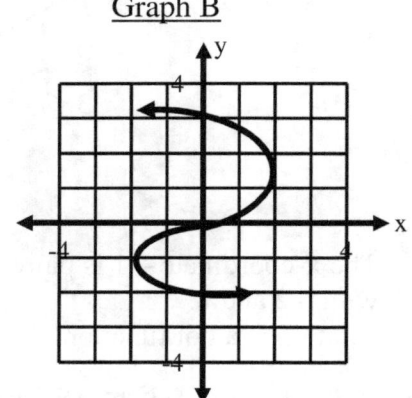

Answers:
For Graph A, any vertical line intersects the graph in exactly one point. Observe that the y-axis intersects the graph at the point (0, 2.5). Graph A passes the Vertical Line Test and each x-coordinate in the graph has exactly one y-coordinate. Therefore Graph A is a function.

For Graph B, a vertical line may intersect the graph in up to three points. Observe that the y-axis intersects the graph at the points (0, –2), (0, 0) and (0, 3). In other words, the x-coordinate x = 0 has three y-coordinates: – 2, 0 and 3. Therefore Graph B does not pass the Vertical Line Test. Graph B is not a function.

✖️ *Helpful Hints* – **IMPORTANT: (1) A vertical line is <u>not</u> a function. (2) The graph of <u>a</u> function <u>cannot</u> <u>be</u> <u>vertical!</u>**

(2.2)

FUNCTION NOTATION F(x)   *(The function symbol F(x) is pronounced "F of x".)*

F(x) means find the y-coordinate when the x-coordinate is given. If F(x) = "a Math expression", let n be an x-coordinate. F(n) means substitute the number n in place of x in the expression and simplify it. The answer is the y-coordinate.

EXAMPLE 7   Let $f(x) = x^3 - 11x + 4$.

(a) Find f(– 3) and interpret the answer.   (b) Find f(2) and interpret the answer.

Answers:

(a) Using the function $f(x) = x^3 - 11x + 4$, substitute – 3 in place of x to get

$f(-3) = (-3)^3 - 11(-3) + 4$

$f(-3) = -27 + 33 + 4$

$f(-3) = 10$. This means that for the x-coordinate x = – 3, the y-coordinate is y = 10.

Thus f(– 3) = 10 is the same as x = – 3, y = 10. This is also the same as the ordered pair (– 3, 10).

(b) Substitute 2 in place of x to get

$f(2) = (2)^3 - 11(2) + 4$

$f(2) = 8 - 22 + 4$

$f(2) = -10$. This means that for the x-coordinate x = 2, the y-coordinate is y = – 10.

Thus f(2) = – 10 is the same as x = 2, y = – 10. This is also the same as the ordered pair (2, – 10).

❊ *Helpful Hints* – (1) Use the correct symbols when working with function notation. For example, do not interchange the symbols f(x) and f(2). The symbol f(x) represents the original expression $f(x) = x^3 - 11x + 4$. The symbol f(2) means substitute 2 for x to get the y-coordinate, –10.   (2) Since the symbol f(x) means "find the y-coordinate", the expression **f(x) = y** indicates that the symbols *f(x)* and *y* may be interchanged in a function, at any time, if it helps you work the problem correctly. In other words:

Function notation		*Equation in x and y*
$f(x) = 2x - 9$	$\Rightarrow$	$y = 2x - 9$.
$f(x) = x^3 - 11x + 4$	$\Rightarrow$	$y = x^3 - 11x + 4$.

EXAMPLE 8   For the function f(x) = 3x – 5, find the x-coordinate if y = – 2.

Answer: Rewrite the function as the equation y = 3x – 5. Then substitute – 2 in place of y to get

$-2 = 3x - 5$

$3 = 3x$

$x = 1$ (The x-coordinate is 1 if the y-coordinate is – 2.)

EXAMPLE 9   The equation $y = x^2 + 4x + 7$ is a function since each x-coordinate has exactly one y-coordinate. Find f(– 1).

Answer: Rewrite the equation as the function $f(x) = x^2 + 4x + 7$. Then substitute – 1 in place of x.

$f(-1) = (-1)^2 + 4(-1) + 7 = 1 - 4 + 7 = 4$.   Therefore f(– 1) = 4.

31

(2.2)

THE DOMAIN AND RANGE OF A GRAPH

Given a graph, use interval notation or inequality notation to describe the domain and range.

To state the **domain**, look at a graph from the left end to the right end and describe the x-coordinates of the graph. Remember that an arrow at the left end of the graph is denoted as "– ∞"; an arrow at the right end of the graph is denoted as "∞".

To state the **range**, look at a graph vertically from the bottom to the top and describe the y-coordinates of the graph. Recall that an arrow at the lowest end of the graph is denoted as "– ∞"; an arrow at the highest end of the graph is denoted as "∞".

EXAMPLE 10   State the domain and range for each graph below.

Graph A

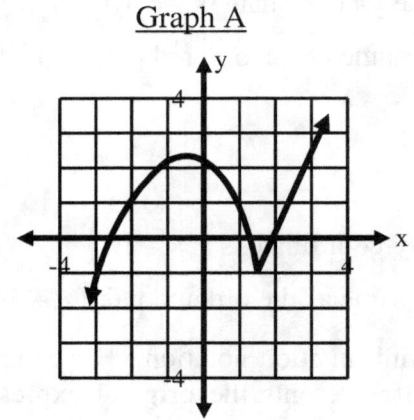

Graph B

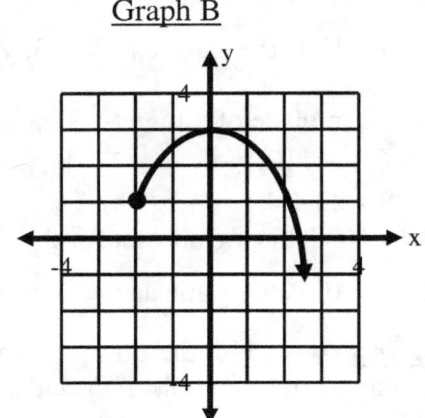

Answers:

(a) For Graph A, the left end of the graph has an arrow which indicates – ∞. The graph continues to the right end which has an arrow to indicate ∞. Therefore the x-coordinates of the graph are from – ∞ to ∞ and the Domain = (– ∞, ∞) = {All real numbers}.

The lowest end of the graph has an arrow indicating – ∞. The highest end of the graph has an arrow which indicates ∞. Thus the y-coordinates of the graph are from – ∞ to ∞ and the Range = (– ∞, ∞) = {All real numbers}.

(b) For Graph B, the point at the left end of the graph has the x-coordinate – 2. The graph continues to the right end which has an arrow to indicate ∞. Therefore the x-coordinates of the graph are from – 2 to ∞ and the Domain = [– 2, ∞) = {x | x ≥ – 2}.

The lowest end of the graph has an arrow indicating – ∞. The highest point of the graph has the y-coordinate 3. Thus the y-coordinates of the graph are from – ∞ to 3 and the Range = (– ∞, 3] = {y | y ≤ 3}.

✠ *Helpful Hints*: **More About Stating the Domain and Range**

If a relation or function is given as an equation or a graph, then the *domain is an interval*, not a list of x-coordinates. The *range is an interval*, not a list of y-coordinates.

EXAMPLE 11   For the relation $y = x^2 + 1$:

(a) The following is a 5-point table of ordered pairs and the graph.

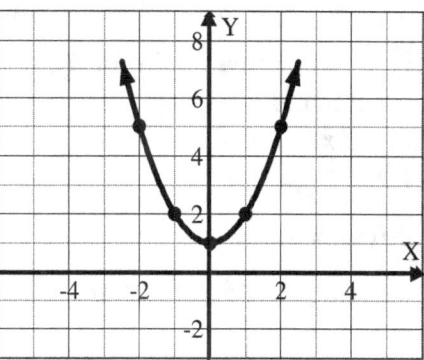

x	y	
-2	5	$y = (-2)^2 + 1 = 4 + 1$
-1	2	$y = (-1)^2 + 1 = 1 + 1$
0	1	$y = (0)^2 + 1 = 0 + 1$
1	2	$y = (1)^2 + 1 = 1 + 1$
2	5	$y = (2)^2 + 1 = 4 + 1$

Observe that the table only shows *5 sample points*. The graph represents **all points** which are on the curve. *For the domain and range, refer to the graph,* not the table of ordered pairs.

(b) For $y = x^2 + 1$, the domain $= (-\infty, \infty) = \{$all real numbers$\}$. The range $= [1, \infty) = \{y \mid y \geq 1\}$.
Both the domain and the range are intervals.

(c) This relation is a function. The graph passes the Vertical Line Test and each x-coordinate has exactly one y-coordinate.

THE DOMAIN FOR THREE TYPES OF FUNCTIONS

Let the function f(x) be given as an equation. To state the domain of f(x), use an inequality or interval notation to describe the x-coordinates which may be used in a table of ordered pairs.

Monomial and Polynomial Functions

The domain $= \{$all real numbers$\} = (-\infty, \infty)$ unless stated otherwise with the function.

EXAMPLE 12   State the domain for $f(x) = x^2 - 1$.

Answer: The function f(x) is a binomial. The domain of f(x) $= \{$all real numbers$\} = (-\infty, \infty)$

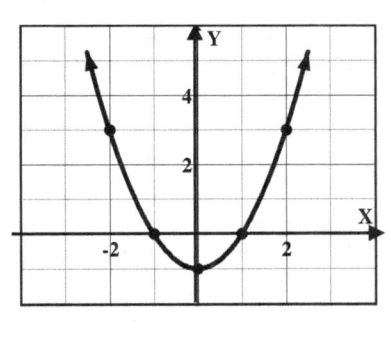

x	$f(x) = x^2 - 1$
⋮	⋮
-2	3
-1	0
0	-1
1	0
2	3
⋮	⋮

(2.2)

✠ **_Helpful Hint_** – The domain "all real numbers" means that in the table of ordered pairs for f(x), we can *substitute <u>any</u> <u>real</u> <u>number</u> for x and get a real number as the y-coordinate.*
This is NOT true for the next two categories.

Square Root Functions

The domain = {x | radicand ≥ 0}. Then solve the inequality for x.

EXAMPLE 13   State the domain for $f(x) = \sqrt{x-1}$
Answer: Radicand ≥ 0

$$x - 1 \geq 0$$
$$x \geq 1$$

The domain of f(x) = {x | x ≥ 1} = [1 , ∞).

Therefore, if we make a table of ordered pairs,
the x-coordinates will start at x = 1.
Also, (1, 0) is the starting point of the graph.

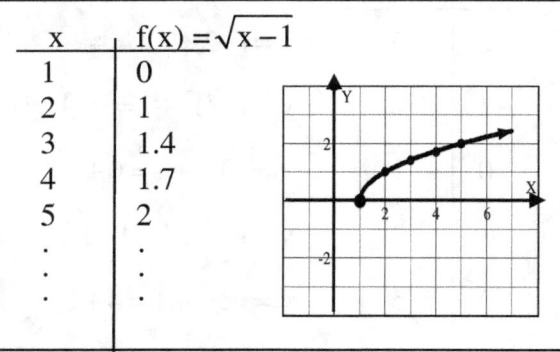

x	$f(x) = \sqrt{x-1}$
1	0
2	1
3	1.4
4	1.7
5	2
⋮	⋮

✠ **_Helpful Hint_** – In Example 13, if we substitute an x-coordinate lower than 1, we will get imaginary numbers for the answer. *<u>Imaginary numbers are **not used** for y-coordinates</u>* in College Algebra. Refer to Section 2.1, Examples 7–9.

Rational Functions   (x in the denominator):

The domain = {x | x ≠ the restricted value(s) of the denominator}

EXAMPLE 14   State the domain $f(x) = \dfrac{2}{x}$.

Answer: Denominator ≠ 0   ⇒   x ≠ 0   ⇒   Zero is the restricted value of the denominator.
The domain of f(x) = {x | x ≠ 0} = (– ∞, 0) ∪ (0, ∞). This means that for x = 0, f(x) is *undefined* and there is no y-coordinate. For any real number other than 0, we can substitute it for x and simplify the result to get the y-coordinate.

x	$f(x) = \dfrac{2}{x}$
⋮	⋮
– 3	– 0.67
– 2	– 1
– 1	– 2
– 0.5	– 4
0	undefined
0.5	4
1	2
2	1
3	0.67
⋮	⋮

�֎ *__Helpful Hint__* – In Example 14, observe that for the restricted value x = 0 there is no y-coordinate. The graph splits into two separate curves. One curve is to the left of x = 0 and the other curve is to the right of x = 0. Refer to Section 2.1, Example 10.

EXAMPLE 15   State the domain for each function:

$$\text{(a)} \ H(x) = \frac{1}{x+3} \qquad \text{(b)} \ F(x) = x^2 + x + 3 \qquad \text{(c)} \ G(x) = \sqrt{x+3}$$

Answers:

(a) $H(x) = \dfrac{1}{x+3}$ is a rational function; the denominator cannot equal zero. The restricted value of the denominator is $x \neq -3$. The domain $= \{x \mid x \neq -3\} = (-\infty, -3) \cup (0, -3)$.

For $x = -3$, the y-coordinate is undefined (i.e. there is no y-coordinate). For any real number other than -3, we can substitute it for x and the result will be a real number as the y-coordinate.

(b) $F(x) = x^2 + x + 3$ is a polynomial function. The domain $= \{\text{all real numbers}\} = (-\infty, \infty)$.

In the x-column of a table of ordered pairs, any real number may be substituted for x and the result will be a real number as the y-coordinate.

(c) $G(x) = \sqrt{x+3}$ is a square root function. Set the radicand ≥ 0 and solve for x.
$$x + 3 \geq 0$$
$$x \geq -3$$
The domain $= \{x \geq -3\} = [-3, \infty)$.

In a table of ordered pairs, the x-coordinates must start at -3 and include numbers greater than -3. The resulting y-coordinates will be real numbers. (Remember, if a number less than -3 is substituted for x, the result is an imaginary number which cannot be used as a y-coordinate.)

EXAMPLE 16   Graph the function $J(x) = -2|x| + 1$. State the domain and the range.

Answers: (Refer to Section 2.1, Example 5.)

x	$J(x) = -2|x| + 1$	(Workspace)
-2	-3	$J(-2) = -2|-2| + 1 = -3$
-1	-1	$J(-1) = -2|-1| + 1 = -1$
0	0	$J(0) = -2|0| + 1 = 0$
1	-1	$J(1) = -2|1| + 1 = -1$
2	-3	$J(2) = -2|2| + 1 = -3$

Graph:

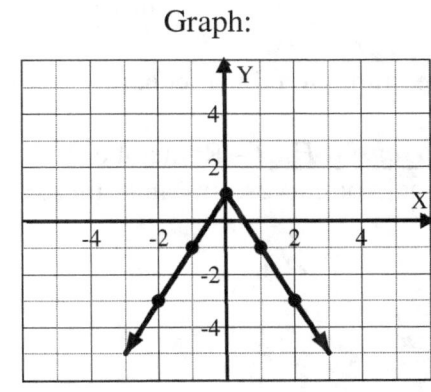

$J(x)$ involves the absolute value of the monomial x.
For $J(x)$, the domain $= (-\infty, \infty) = \{\text{all real numbers}\}$. The range $= (-\infty, 1]$.

(2.2)

EXAMPLE 17   Graph the function $P(x) = x^3 + 1$. State the domain and the range.

Answers: (Refer to Section 2.1, Example 6.)

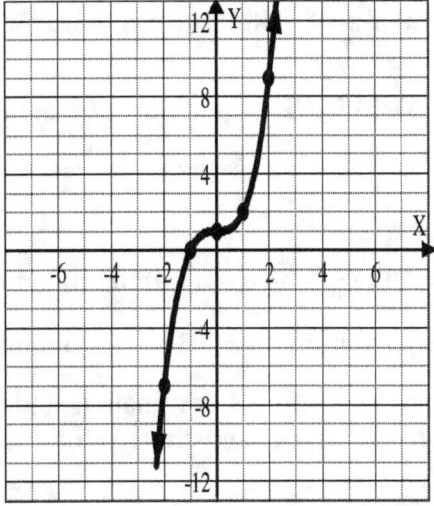

x	$P(x) = x^3 + 1$	(Workspace)
– 2	– 7	$P(-2) = (-2)^3 + 1 = -7$
– 1	0	$P(-1) = (-1)^3 + 1 = 0$
0	1	$P(0) = (0)^3 + 1 = 1$
1	2	$P(1) = (1)^3 + 1 = 2$
2	9	$P(2) = (2)^3 + 1 = 9$

$P(x) = x^3 + 1$ is a polynomial function.
The domain = {all real numbers} = $(-\infty, \infty)$.
The range = {all real numbers} = $(-\infty, \infty)$.

2.3   LINEAR FUNCTIONS AND EQUATIONS

LINEAR EQUATIONS IN TWO VARIABLES

The **standard form** of a linear equation in two variables is **Ax + By = C**, where A, B and C are the constants of the equation. The linear equation is a first-degree equation. It has no exponents. Also, there is no variable in the denominator of a fraction.

EXAMPLE 1   The equation $3x + 5y = 2$ is linear. The following are <u>not</u> linear equations:

$3x^2 + 5y = 2$ (It has an exponent.) $\dfrac{3}{x} + 5y = 2$ (It has x in the denominator.)

The graph of a linear equation is a **straight line**.

� *Helpful Hint* – Remember that "straight line" does *not* mean vertical line. A straight line has no corners or curves. In Math, there are four straight lines:

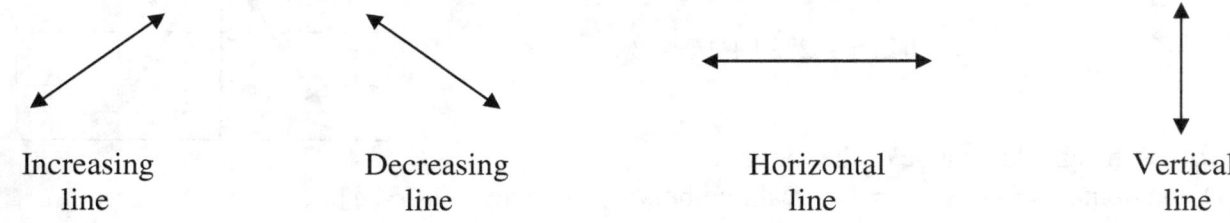

| Increasing | Decreasing | Horizontal | Vertical |
| line | line | line | line |

(2.3)

THE INTERCEPTS

The **x-intercept** is the point where the graph intersects the
x-axis. For the graph at the right, the x-intercept is -2 or $(-2, 0)$.
The **y-intercept** is the point where the graph intersects the
y-axis. For the graph at the right, the y-intercept is 3 or $(0, 3)$.

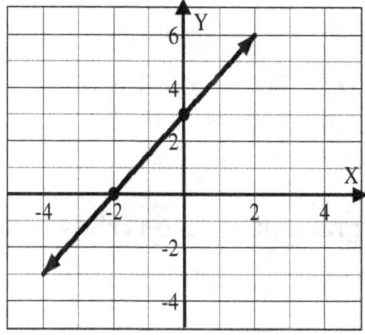

Given a linear equation in x and y:
The x-intercept has the format $(x, 0)$. Let $y = 0$ and find the
x-coordinate.

The y-intercept has the format $(0, y)$. Let $x = 0$ and find the
y-coordinate.

Graphing Linear Equations: The Intercepts Method
To graph the linear equation, find the x-intercept and the y-intercept. Place a point at both intercepts in
the rectangular system and draw the line through the points.

EXAMPLE 2   Find the intercepts and graph $2x + y = 4$.

Answer: Make an **Intercepts Table** by placing a zero in the
x-column of the first line and a zero in the y-column of the
second line. Find the remaining coordinates. Locate both
intercepts in the rectangular system. Draw the line through
points. The graph of $2x + y = 4$ is shown at the right.

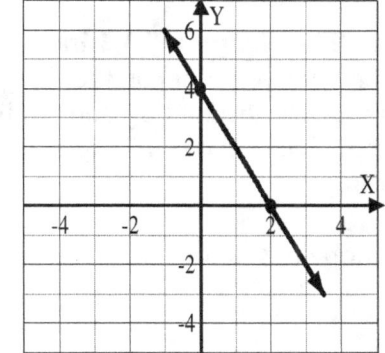

Intercepts Table   (Workspace)

x	y
0	4
2	**0**

$x = 0$

$2(0) + y = 4$

$0 + y = 4$

$y = 4$

The y-intercept is $(0, 4)$.

$y = 0$

$2x + 0 = 4$

$2x = 4$

$x = 2$

The x-intercept is $(2, 0)$.

✳ *Helpful Hint* – If a linear equation has *two variables*, then the graph is a slanted line. Most linear
equations have two variables. Also, a linear equation in two variables is a **linear function**.

SPECIAL CASES – LINEAR EQUATIONS IN ONE VARIABLE

The equation **y = b** is a also linear function where each y-coordinate is the
same number **b**. The y-intercept is $(0, b)$. The graph is a horizontal line.

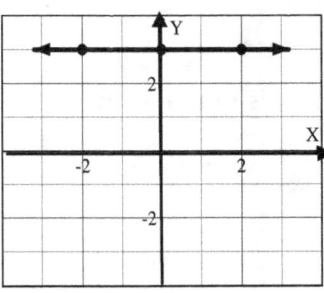

EXAMPLE 3   Graph $y = 3$.

Answer: For each point, the y-coordinate is 3.
The y-intercept is $(0, 3)$ and the graph is a horizontal
line through the y-intercept.
The graph of $y = 3$ is shown at the right.

x	y
-2	3
0	3
2	3

(2.3)

The linear equation **x = c** is *not* a function since each x-coordinate is the same number **c**. The x-intercept is (**c**, 0). The graph is a vertical line.

EXAMPLE 4   Graph x = 2.6 .

Answer: The x-coordinate for each point is 2.6 and the x-intercept is (2.6, 0). The graph is a vertical line through the x-intercept.

The graph of x = 2.6 is shown at the right.

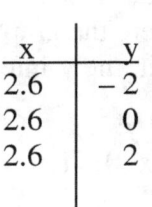

x	y
2.6	− 2
2.6	0
2.6	2

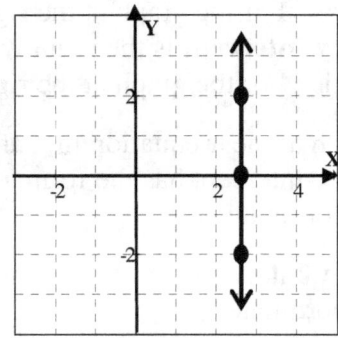

SLOPE OF A LINE

The slope is a fraction or integer which indicates the steepness of a line. The symbol for slope is the lowercase letter **m**. There are three ways to determine the slope of a line depending on the information given in the problem.

Given The Graph Of A Line:
Select any two points on the line. Start at the lower point and make a right angle, straight up and straight across to the upper point. The vertical side of the right angle is the rise, and the horizontal side is the run.

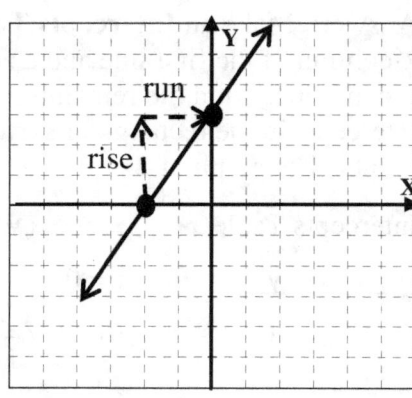

The slope of the line is

$$slope\ \boldsymbol{m} = \frac{rise}{run}$$

EXAMPLE 5   For the graph at the right, slope m = $\dfrac{rise}{run} = \dfrac{3}{2}$.

EXAMPLE 6   For the graph at the right, find the slope of the line.

Answer: Select any two points on the line. Start at the lower point and make a right angle, straight up and straight across (left) to the upper point.

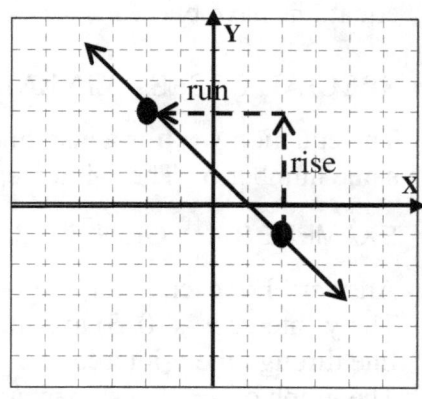

For this graph, slope m = $\dfrac{rise}{run} = \dfrac{4}{-4} = -1$.

Given Two Points On The Line, (x_1, y_1) and (x_2, y_2):

$$\text{slope } \mathbf{m} = \frac{\text{rise}}{\text{run}} = \frac{y_2 - y_1}{x_2 - x_1}$$

EXAMPLE 7   Find the slope of the line that passes through $(1, -3)$ and $(5, -6)$.

Answer: $x_1 = 1$, $y_1 = -3$, $x_2 = 5$ and $y_2 = -6$. Slope $m = \dfrac{-6-(-3)}{5-1} = \dfrac{-3}{4} = -\dfrac{3}{4}$

Note: The third method to determine the slope of a line is presented in Section 2.4 of this chapter.

Sign Of The Slope

The sign of the slope indicates the direction of the line.

Positive slope m: The line is slanted, pointing up at the right.

Negative slope m: The line is slanted, pointing down at the right.

Slope m = 0: The line is horizontal.

Slope m = undefined: The line is vertical.

THE SLOPE OF PARALLEL LINES AND PERPENDICULAR LINES

Parallel lines do not intersect, no matter how much they are extended.
The Math symbol for parallel is ‖ .

Perpendicular lines intersect at right angles (90°).
The Math symbol for perpendicular is ⊥ .

Let $\mathbf{L_1}$ and $\mathbf{L_2}$ represent two lines; $\mathbf{m_1}$ is the slope of line $\mathbf{L_1}$ and $\mathbf{m_2}$ is the slope of line $\mathbf{L_2}$.

- If two lines are parallel, then their slopes are equal. In symbols, $\mathbf{L_1} \,\|\, \mathbf{L_2} \;\Leftrightarrow\; \mathbf{m_2 = m_1.}$
- If two lines are perpendicular, then the slope of the second line is the negative reciprocal of the

    first slope. In symbols, $\mathbf{L_1} \perp \mathbf{L_2} \;\Leftrightarrow\; \mathbf{m_2} = -\dfrac{1}{m_1}$.

    (To find the slope m_2 for the perpendicular line, change the sign and invert the first slope m_1.)

EXAMPLE 8   Given the slope of line L_1 is $m_1 = \dfrac{2}{5}$:

(a) The slope of any line parallel to line L_1 is $m_2 = \dfrac{2}{5}$.

(b) The slope of any line perpendicular to line L_1 is $m_2 = -\dfrac{5}{2}$.

(2.3)

EXAMPLE 9 L_1 is a line through the points (3, 2) and (4, – 1). (a) Find the slope of line L_2 which is parallel to the line L_1. (b) Find the slope of line L_3 which is perpendicular to the line L_1.

Answers:

(a) For line L_1, slope $m_1 = \frac{y_2 - y_1}{x_2 - x_1} = \frac{-1-2}{4-3} = \frac{-3}{1} = -3$. Since $L_1 \parallel L_2$, slope $m_2 = -3$.

(b) For line L_1, slope $m_1 = -3$. Since $L_1 \perp L_3$, slope $m_3 = \frac{1}{3}$.

2.4 LINEAR FUNCTIONS: SLOPE-INTERCEPT FORM

Given Ax + By = C, solve for y to get the **slope-intercept form: y = mx + b**, a linear function where **m** is the slope and **b** is the y-intercept. Also, f(x) = y. In function notation, we write **f(x) = mx + b**.

EXAMPLE 1 State the slope and y-intercept for the following linear functions:

(a) $y = 4x - 1$ (b) $f(x) = \frac{3}{2}x + 2$ (c) $f(x) = -2x$ (d) $y = -2$

Answers: Each equation is already solved for y.

(a) For $y = 4x - 1$ or $f(x) = 4x - 1$, slope m = 4 and y-intercept b = – 1.

(b) For $f(x) = \frac{3}{2}x + 2$ or $y = \frac{3}{2}x + 2$, slope $m = \frac{3}{2}$ and y-intercept b = 2.

(c) For $f(x) = -2x$ or $y = -2x$, slope m = – 2. The y-intercept b = 0 (since the second term is missing).

(d) For $y = -2$ or $f(x) = -2$, slope m = 0 since the *x* term is missing. The y-intercept b = – 2.

In Section 2.1, we graphed a linear equation by using a table of ordered pairs. In Section 2.3, we graphed a linear equation by using the x-intercept and the y-intercept. The most commonly used procedure for graphing a linear equation or function is the **Slope-Intercept Method**. This procedure is used to graph a linear equation *without using a table of ordered pairs*. The following are the steps:

Graphing Linear Functions: The Slope-Intercept Method

- Solve the equation for y.
- Write the slope and y-intercept.
- Place a point at the y-intercept (which is on the y-axis).
- Start at the y-intercept and count the slope $m = \frac{\text{rise}}{\text{run}}$.
  Make a right angle and place the second point.
- Draw a straight line through the points.

Helpful Hints – (1) The Slope-Intercept Method is also the *third method for finding the slope of a line, given an equation in two variables.* (2) For graphing, the slope m should be written as a fraction. (3) The y-intercept may be written as an ordered pair (0, b) since the y-intercept is a point on the y-axis. (4) To locate more than two points of the graph, start from the second point and count the slope again to get a third point of the graph, etc.

EXAMPLE 2   State the slope and y-intercept. Use this information to draw the graph for y = 3x – 2.

Answer: (a) This equation is already solved for y. (b) For y = 3x – 2, slope m = 3 = $\dfrac{3}{1}$ = $\dfrac{\text{rise}}{\text{run}}$.

The y-intercept is b = – 2 which is the ordered pair (0, – 2). (c) Place the first point on the y-axis at
(0, – 2). (d) From the first point, count the slope by making a right angle with a vertical rise of 3
spaces and a horizontal run of 1 space to the right. Place the second point. (e) Neatly draw a straight
line through the points.

Graph of y = 3x – 2

Steps (c) – (d)

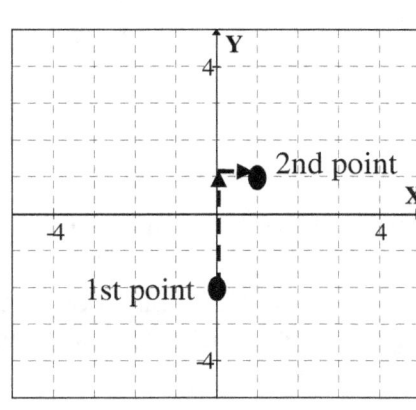

Step (e)

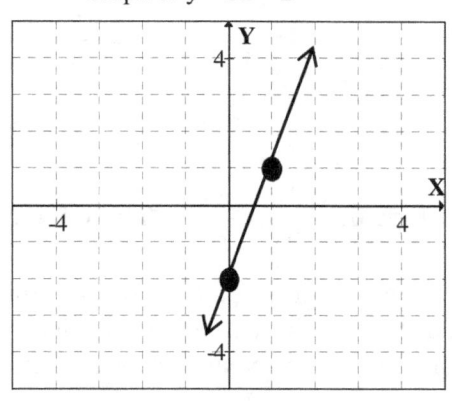

EXAMPLE 3   Find the slope and y-intercept. Then draw the graph for 3x + 2y = 2.

Answer: (a) Solve for y.

$$3x + 2y = 2$$
$$\underline{-\,3x \qquad\quad -\,3x} \qquad \text{Subtract 3x from both sides of the equation.}$$
$$2y = -\,3x + 2$$
$$\frac{2y}{2} = \frac{-\,3x}{2} + \frac{2}{2} \qquad \text{Divide each term by 2.}$$
$$y = -\frac{3}{2}x + 1 \quad \text{or} \quad f(x) = -\frac{3}{2}x + 1$$

(b) The slope m = $\dfrac{-\,3}{2}$ = $\dfrac{\text{rise}}{\text{run}}$. The y-intercept is b = 1 which is the ordered pair (0, 1).

(c) Place the first point on the y-axis at (0, 1). (d) From the first point, count the slope by making a
right angle with a vertical rise of – 3 (which is 3 spaces *downward*) and a horizontal run of 2 spaces to
the right. Place the second point. (e) Neatly draw a straight line through the points.

Graph of 3x + 2y = 2

Steps (c) – (d)

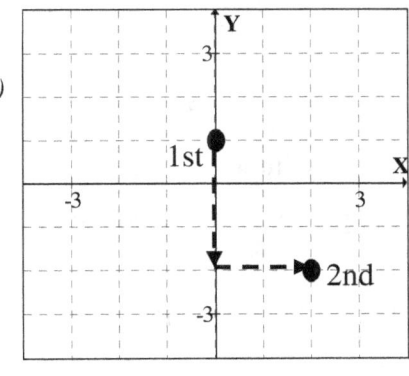

Step (e)

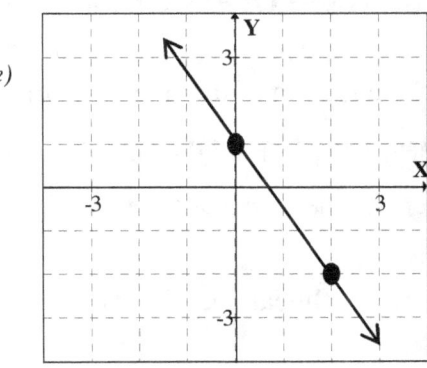

(2.4)

EXAMPLE 4   Find the slope and y-intercept. Then draw the graph for x – 3y = 0.

Answer: (a) Solve for y.

$$x - 3y = 0$$
$$\underline{-x \qquad\quad -x}$$ Subtract x from both sides of the equation.

$$-3y = -x$$

$$\frac{-3y}{-3} = \frac{-x}{-3}$$ Divide each term by – 3.

$$y = \frac{x}{3}$$

$$y = \frac{1}{3}x \quad \text{or} \quad f(x) = \frac{1}{3}x$$

(b) The slope $m = \dfrac{1}{3} = \dfrac{\text{rise}}{\text{run}}$. The y-intercept is b = 0 which is the ordered pair (0, 0).

(c) Place the first point on the y-axis at the origin (0, 0). (d) From the first point, count the slope by making a right angle with a vertical rise of 1 space and a horizontal run of 3 spaces to the right. Place the second point. (e) Neatly draw a straight line through the points.

Graph of x – 3y = 0

Steps (c) – (d)
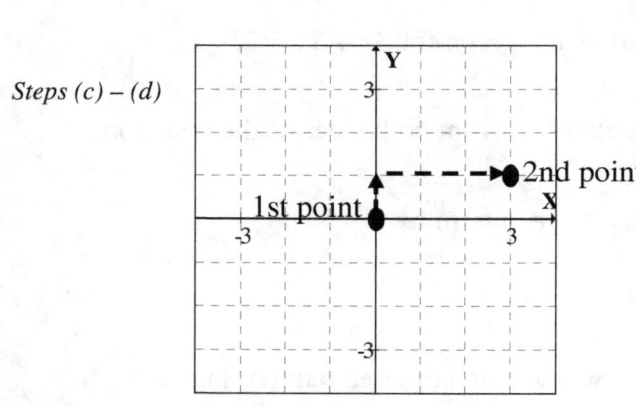

Step (e)
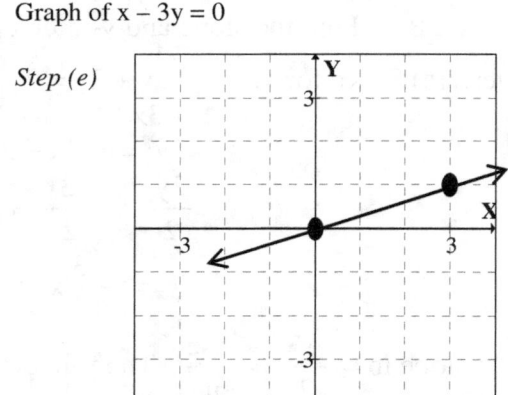

2.5   FINDING THE EQUATION OF A LINE

USING THE SLOPE AND Y-INTERCEPT

Given the slope *m* and y-intercept *b* of a line, you may use the slope-intercept format **y = mx + b** to write the equation of the line. The linear function for the line is **f(x) = mx + b**.

EXAMPLE 1   Find the equation for the line having slope 3 and y-intercept – 5.

Answer: Slope m = 3 and y-intercept b = – 5. Using the format y = mx + b, the equation for the line is y = 3x + (– 5).

Therefore the linear equation is y = 3x – 5.   The linear function for the line is f(x) = 3x – 5.

USING THE SLOPE AND ANY POINT

Given slope m and any point (x_1 , y_1), use the **Formula for the Equation of a Line**:

$$y - y_1 = m(x - x_1)$$

This formula is also called the **Point-Slope Formula**. Substitute for m, x_1 and y_1. Then solve the equation for y.

EXAMPLE 2   Find the equation for the line through (4, 2) whose slope is 3.

Answer: Slope m = 3, x_1 = 4 and y_1 = 2. Substitute these values into the above formula and solve the equation for y.

$$y - y_1 = m(x - x_1)$$
$$y - 2 = 3(x - 4) \qquad \text{Substitute for } y_1, \, m \text{ and } x_1.$$
$$y - 2 = 3x - 12 \qquad \text{Simplify the parentheses.}$$
$$\underline{+2 \qquad +2} \qquad \text{Solve for } y.$$
$$y = 3x - 10 \qquad \text{Result}$$

The linear equation is y = 3x – 10.   The linear function for the line is f(x) = 3x – 10.

EXAMPLE 3   Find the equation for the line through (– 3, 5) and (2, – 1).

Answer: To use the Formula for the Equation of a Line, we must have the slope of the line and a point on the line. In this example, we are given two points on the line. First, find the slope by using the slope formula. For the given points, x_1 = – 3, y_1 = 5, x_2 = 2 and y_2 = – 1.

$$\text{slope formula } \; m = \frac{y_2 - y_1}{x_2 - x_1} = \frac{-1-5}{2-(-3)} = \frac{-6}{2+3} \quad \rightarrow \quad m = \frac{-6}{5}$$

The second step is to substitute for m, x_1 and y_1 in the Formula for the Equation of a Line and solve for y.

$$y - y_1 = m(x - x_1)$$
$$y - 5 = -\frac{6}{5}(x - (-3)) \qquad \text{Substitute for } y_1, \, m \text{ and } x_1.$$
$$y - 5 = -\frac{6}{5}(x + 3) \qquad \text{Simplify the parentheses.}$$
$$y - 5 = -\frac{6}{5}x - \frac{18}{5} \qquad \text{Solve for } y.$$
$$\underline{+5 \qquad\qquad +5}$$
$$y = -\frac{6}{5}x - \frac{18}{5} + \frac{5}{1}\left(\frac{5}{5}\right) \qquad \text{Set up a common denominator of 5.}$$
$$y = -\frac{6}{5}x - \frac{18}{5} + \frac{25}{5} \qquad \text{Simplify the fractions.}$$
$$\text{Result} \qquad y = -\frac{6}{5}x + \frac{7}{5} \qquad \text{Linear equation in } x \text{ and } y$$

Also, the linear function for the line through (– 3, 5) and (2, – 1) is $f(x) = -\frac{6}{5}x + \frac{7}{5}$.

(2.5)

✠ *Helpful Hint* – To write the equation in **standard form, Ax + By = C**, eliminate the fractions. Rearrange the equation with the variable terms on the left side of the equation and constants on the right side.

$$y = -\frac{6}{5}x + \frac{7}{5}$$

$$y(5) = -\frac{6}{5}x(5) + \frac{7}{5}(5) \qquad \text{Multiply each term times the LCD 5.}$$

$$y(5) = -\frac{6}{5}x(5) + \frac{7}{5}(5) \qquad \text{Cancel out each denominator.}$$

$$5y = -6x + 7 \qquad \text{Simplify each term.}$$
$$+6x \qquad +6x \qquad \text{Add 6x to each side of the equation.}$$

Answer: $6x + 5y = 7$ This result is the standard form for the equation $y = -\frac{6}{5}x + \frac{7}{5}$.

A VERTICAL LINE - - UNDEFINED SLOPE

Recall that the slope of a vertical line is undefined (meaning there is no answer for the slope). Therefore the Formula for the Equation of a Line *cannot* be used for a vertical line. Use the formula **x = c** where **c** is the given x-coordinate.

EXAMPLE 4   Find the equation of the line through $(2, 7)$ and $(2, -4)$.

Answer: In this example, we are given two points on the line. First find the slope by using the slope formula. For the given points, $x_1 = 2$, $y_1 = 7$, $x_2 = 2$ and $y_2 = -4$.

$$\text{slope formula} \quad m = \frac{y_2 - y_1}{x_2 - x_1} = \frac{-4 - 7}{2 - 2} = \frac{-11}{0} = \text{undefined} \quad (\text{or } no\ solution)$$

The undefined slope indicates that the line through $(2, 7)$ and $(2, -4)$ is vertical. The formula for a vertical line is $x = c$. Observe that both points have the same x-coordinate 2, which is the value of c. The equation of the line is $x = 2$.

APPLICATIONS

EXAMPLE 5   Tip-Top Discount Store sells a $48 sweater for $30 and a $108 jacket for $70. Write a linear function that expresses discount price y in terms of the original price x.

Answer: Write the numerical data as two ordered pairs $(48, 30)$ and $(108, 70)$. Find the slope.

$$\text{slope formula} \quad m = \frac{y_2 - y_1}{x_2 - x_1} = \frac{70 - 30}{108 - 48} = \frac{40}{60} \quad \rightarrow \quad m = \frac{2}{3}$$

The formula for the linear equation is $y - y_1 = m(x - x_1)$

$$y - 30 = \frac{2}{3}(x - 48)$$

$$y - 30 = \frac{2}{3}x - 32$$

The linear equation is $y = \frac{2}{3}x - 2$, and the linear function is $f(x) = \frac{2}{3}x - 2$.

EXAMPLE 6   Using the graph at the right, identify each of the following items:
(a) slope m (b) y-intercept b
(c) the equation of the line   (d) the x-intercept

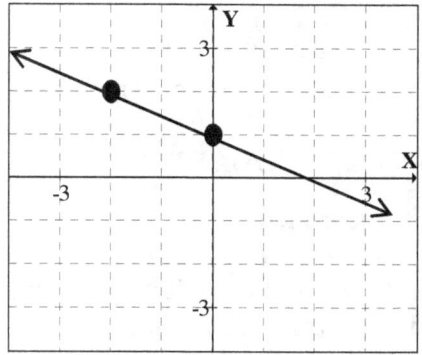

Answers:
(a) Starting at the lower point (0, 1), make a right angle to the upper point (– 2, 2). The rise is 1 space and the run is 2 spaces *left*. The slope m $= \dfrac{\text{rise}}{\text{run}} = \dfrac{1}{-2} \rightarrow m = -\dfrac{1}{2}$

(b) The y-intercept is the point (0, 1), which is on the y-axis. Thus y-intercept $b = 1$.

(c) Based on the answers for the slope and y-intercept, we can easily use the slope-intercept format $y = mx + b$. The equation of the line is $y = -\dfrac{1}{2}x + 1$. In function notation, $f(x) = -\dfrac{1}{2}x + 1$.

(d) The x-intercept is the point (2, 0), which is on the x-axis.

2.6   LINEAR INEQUALITIES IN TWO VARIABLES

Whenever we graph a line, each side of the line is called a **half-plane**.

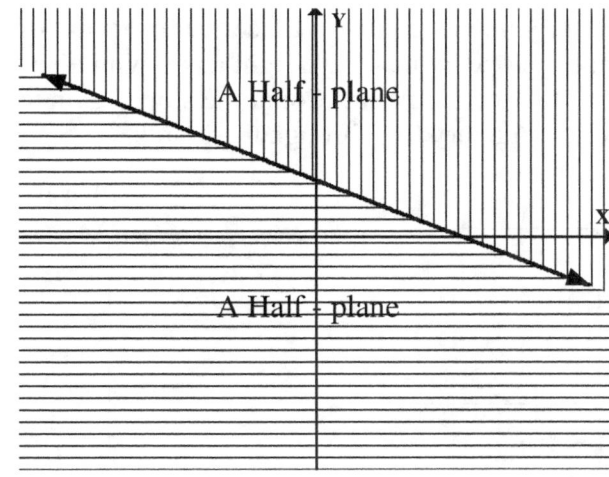

The solution set of a **linear inequality in two variables** is the half-plane containing all points which make the inequality true. (Only one side of the line is the "true" half-plane.)

EXAMPLE 1   Determine whether (0, 0) is in the solution set for the inequality y > 2x – 3.

Answer: Substitute the first 0 for x and the second 0 for y. Then simplify.

    0 > 2(0) – 3
    0 > 0 – 3
    0 > – 3 is True. Therefore (0, 0) is in the solution set for y > 2x – 3.

(2.6)

EXAMPLE 2   Determine whether $(2, -2)$ is a solution for the inequality $y > 2x - 3$.

Answer: Substitute 2 for x and -2 for y. Then simplify.

$-2 > 2(2) - 3$
$-2 > 4 - 3$
$-2 > 1$ is False. Therefore $(2, -2)$ is *not* a solution for $y > 2x - 3$.

LINEAR INEQUALITY WITH $\leq$, OR, $\geq$

For a linear inequality that contains the $\leq$ symbol or the $\geq$ symbol, the solution set is the *graph of the solid boundary line and the shading of one half-plane* containing all points which make the inequality true. (The solid boundary line is part of the solution set.) Therefore, graph the solid boundary line and shade the "true" half-plane.

EXAMPLE 3   Find the slope and y-intercept. Then draw the graph for $x - y \geq 2$

Answer: (a) Solve for y.

$$x - y \geq 2$$
$$\underline{ -x -x}\qquad \text{Subtract x from both sides of the inequality.}$$
$$-y \geq -x + 2$$
$$\frac{-y}{-1} \geq \frac{-x}{-1} + \frac{2}{-1}\qquad \text{Divide each term by } -1 \text{ and}$$
$$ \text{change the inequality symbol.}$$
$$y \leq x - 2\qquad \text{Result}$$

(b) The slope $m = \dfrac{1}{1} = \dfrac{\text{rise}}{\text{run}}$. The y-intercept is $b = -2$ which is the ordered pair $(0, -2)$. The $\leq$ symbol indicates that the boundary line is solid. (c) Place the first point on the y-axis at $(0, -2)$. (d) From the first point, count the slope by making a right angle with a vertical rise of 1 space and a horizontal run of 1 space. Place the second point. Count the slope again to get a third point on the x-axis. (e) Neatly draw the solid boundary line through the points.

Steps (c) – (d)

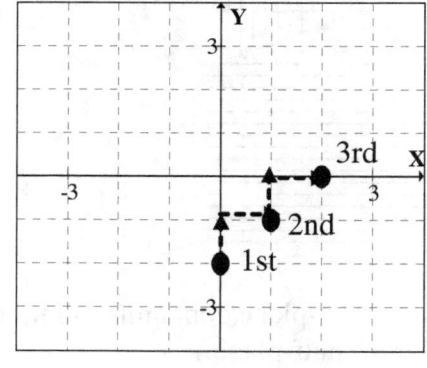

Boundary line for $x - y \geq 2$

Steps (e) – (f)

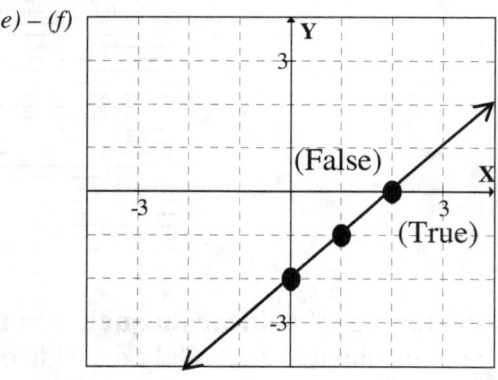

(f) For the inequality, choose a **test point** that is not on the line, such as $(0, 0)$. Substitute the first 0 for x and the second 0 for y in the inequality $y \leq x - 2$ and determine whether the result is true for false: $0 \leq 0 - 2 \rightarrow 0 \leq -2$ is false. Therefore $(0, 0)$ is in the "false" half-plane and the other side of the boundary line is the "true" half-plane. (g) On the graph, shade the "true" half-plane. (The final graph is at the top of the next page.)

46

(2.6)

At the right is the graph of the solution
set for the inequality x – y ≥ 2.

Graph of x – y ≥ 2

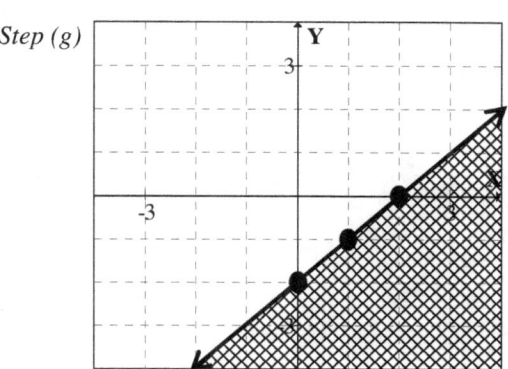

Step (g)

LINEAR INEQUALITY WITH < , OR, >

For a linear inequality that contains the < symbol or the > symbol, the solution set is the *graph of the shaded half-plane* containing all points which make the inequality true. (The *dotted* boundary line is **not** part of the solution set.) Therefore, graph the <u>dotted</u> boundary line and shade the "true" half-plane.

EXAMPLE 4   Find the slope and y-intercept. Then draw the graph for 2x + y < 2.

Answer: (a) Solve for y. $2x + y < 2$

$$-2x \qquad\qquad -2x$$ Subtract 2x from both sides of the equation.

$$y < -2x + 2$$ Result

(b) The slope m = $\dfrac{-2}{1}$ = $\dfrac{\text{rise}}{\text{run}}$. The y-intercept is b = 2 which is the ordered pair (0, 2). The < symbol
indicates that the boundary line is not part of the solution set. The boundary line will be dotted.
(c) Place the first point on the y-axis at (0, 2). (d) From the first point, count the slope by making a
right angle with a vertical rise of – 2 (which is 2 spaces downward) and a horizontal run of 1 space to
the right. Place the second point. (e) Neatly draw a dotted boundary line through the points.

Boundary line for 2x + y < 2

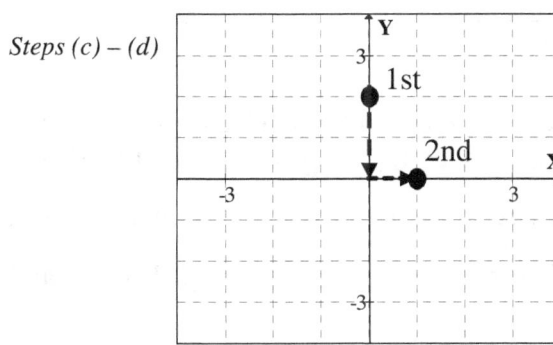

Steps (c) – (d)

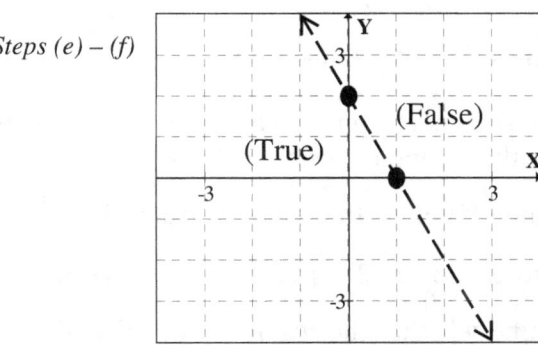

Steps (e) – (f)

(f) For the inequality, choose a *test point* that is not on the line, such as (0, 0). Substitute the first 0 for
x and the second 0 for y in the inequality y < – 2x + 2 and determine whether the result is true or false:
$0 < -2(0) + 2 \;\rightarrow\; 0 \le 0 + 2 \;\rightarrow\; 0 \le 2$ is true. Therefore (0, 0) is in the "true" half-plane and

(2.6)

the other side of the boundary line is the "false" half-plane. (g) On the graph, shade the true half-plane.

Graph of $2x + y < 2$

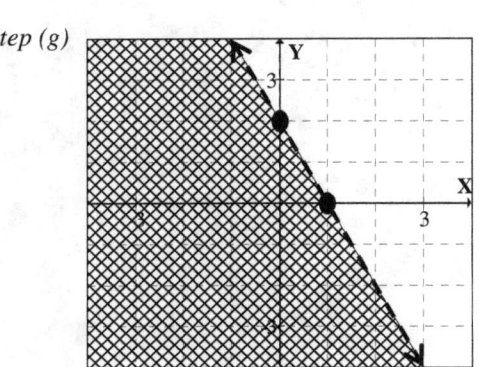

Step (g)

EXAMPLE 5   The graph below is the solution set for which of the following linear inequalities?

[1] $y < \dfrac{1}{2}x + 1$ [2] $y \le \dfrac{1}{2}x + 1$

[3] $y > \dfrac{1}{2}x + 1$ [4] $y \ge \dfrac{1}{2}x + 1$

Answer:

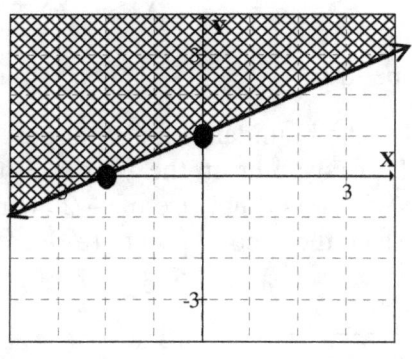

(a) In the graph, the boundary line is solid which indicates that the inequality symbol is either $\le$ or $\ge$. Therefore inequalities [1] and [3] are eliminated. (b) In the graph, y-intercept b = 1 and the slope = $\dfrac{\text{rise}}{\text{run}} = \dfrac{1}{2}$. For both inequalities [2] and [4], slope m = $\dfrac{1}{2}$ and y-intercept b = 1.

(c) The shaded region indicates the "true" half-plane. Choose a *test point* in the shaded region which is not on the boundary line, such as (– 2 , 1). Substitute – 2 for x and 1 for y in inequality [2] and determine whether the result is true: $1 \le \dfrac{1}{2}(-2) + 1 \ \rightarrow \ 1 \le -1 + 1 \ \rightarrow \ 1 \le 0$ which is false.

Therefore choice [2] is eliminated. (d) Now substitute – 2 for x and 1 for y in inequality [4] and determine whether the result is true: $1 \ge \dfrac{1}{2}(-2) + 1 \ \rightarrow \ 1 \ge -1 + 1 \ \rightarrow \ 1 \ge 0$ which is true. The

correct answer is inequality [4]. Therefore the graph is the solution set for $y \ge \dfrac{1}{2}x + 1$.

Chapter 3   Polynomials and Equations

3.1   PROPERTIES OF EXPONENTS

EXPONENTS

The expression 2^3 is read "2 to the 3rd power". The number 2 is the **base**. The **exponent** is 3 (the number at the upper right of 2). For a variable expression, the numerical factor in front of the variable is called the **coefficient**. In the expression $3x^2$, 3 is the coefficient, x is the base, and the exponent is 2.

Definition of Exponent

An exponent means repeatedly multiply the base number times itself. Therefore:

$$4^2 = 4 \cdot 4 = 16 \qquad\qquad 3^3 = 3 \cdot 3 \cdot 3 = 27$$

PROPERTY 1: $\boxed{X^0 = 1 \ \text{and} \ X^1 = X.}$

EXAMPLE 1   (a) $2^3 = 2 \cdot 2 \cdot 2 = 8$ (b) $(-2)^3 = (-2)(-2)(-2) = -8$

(c) $(-2yz)^0 = 1$ (d) $5^1 = 5$ (e) $x^1 y^1 = xy$

PROPERTY 2 – **Product Rule**

When multiplying, add the exponents of each variable: $\boxed{X^M X^N = X^{M+N}}$

EXAMPLE 2   (a) $x^{15} x^{10} = x^{25}$ (b) $(a^4 b^5)(ab^7) = a^5 b^{12}$

PROPERTY 3 – **Basic Rule For Multiplication**

$\boxed{\text{Multiply the coefficients and add the exponents for each variable.}}$

EXAMPLE 3   (a) $(3x^2)(7x^3) = 21x^5$ (b) $(5a^2 b^3)(6ab^4) = 30a^3 b^7$

(c) $(xy^2)(-2x^2 y^3)(-4xy^3) = 8x^4 y^8$

PROPERTY 4 – **Quotient Rules**

When dividing, subtract the exponents of each variable. ("Large minus Small Exponent" Method)

$$\boxed{\frac{X^M}{X^N} = X^{M-N} \ \text{IF} \ M > N \ \text{and} \ \frac{X^M}{X^N} = \frac{1}{X^{N-M}} \ \text{IF} \ N > M}$$

EXAMPLE 4   (a) $\dfrac{x^{11}}{x^5} = x^{11-5} = x^6$ (b) $\dfrac{x^8}{x^{12}} = \dfrac{1}{x^{12-8}} = \dfrac{1}{x^4}$

(3.1)

PROPERTY 5 – **Basic Rule For Division**

Divide the coefficients and subtract the exponents for each variable.

EXAMPLE 5   (a) $\dfrac{10x^4}{2x^2} = 5x^{4-2} = 5x^2$ (b) $\dfrac{3a^3b^8}{12ab^5} = \dfrac{a^{3-1}b^{8-5}}{4} = \dfrac{a^2b^3}{4}$

(c) $\dfrac{70x^5y^2}{10x^3y^8} = \dfrac{7x^{5-3}}{y^{8-2}} = \dfrac{7x^2}{y^6}$

PROPERTY 6 – **The Negative Exponent Rule**

A **negative exponent** means invert the base and use the positive exponent:

$$X^{-N} = \dfrac{1}{X^N}$$

EXAMPLE 6   (a) $x^{-1} = \dfrac{1}{x}$ (b) $2^{-3} = \dfrac{1}{2^3} = \dfrac{1}{8}$

(c) $\left(\dfrac{3}{4}\right)^{-2} = \left(\dfrac{4}{3}\right)^2 = \left(\dfrac{4}{3}\right)\left(\dfrac{4}{3}\right) = \dfrac{16}{9}$

PROPERTY 7 – **Exponent Outside the Parentheses**

Multiply the outside exponent times each inside exponent:

$$(X^M)^N = X^{MN} \qquad (XY)^M = X^M Y^M \qquad \left(\dfrac{X}{Y}\right)^M = \dfrac{X^M}{Y^M}$$

EXAMPLE 7   (a) $(x^5)^7 = x^{35}$ (b) $(2x^3y^5)^3 = 2^3 x^9 y^{15} = 8x^9y^{15}$

(c) $\left(\dfrac{x^2}{y^3}\right)^5 = \dfrac{x^{10}}{y^{15}}$ (d) $\left(\dfrac{x^2}{3}\right)^{-3} = \left(\dfrac{3}{x^2}\right)^3 = \dfrac{27}{x^6}$

MORE EXAMPLES: USING THE PROPERTIES OF EXPONENTS

EXAMPLE 8 $(2x^4y^3z^2)^3(5x^7y^2z)^2 = (8x^{12}y^9z^6)(25x^{14}y^4z^2) = 200x^{26}y^{13}z^8$

EXAMPLE 9 $\left(\dfrac{24a^5b^{-7}c^6}{8ab^{-5}c^{-3}}\right)^{-2} = \left(\dfrac{24a^5b^5c^6c^3}{8ab^7}\right)^{-2} = \left(\dfrac{3a^4c^9}{b^2}\right)^{-2} = \left(\dfrac{b^2}{3a^4c^9}\right)^2 = \dfrac{b^4}{9a^8c^{18}}$

EXAMPLE 10 $\dfrac{(2x^2y^{-3}z)^3(5xy^2)}{(3xy^2z^{-10})^{-1}} = \dfrac{(8x^6y^{-9}z^3)(5xy^2)}{3^{-1}x^{-1}y^{-2}z^{10}}$

$= \dfrac{(8x^6z^3)(5xy^2)(3xy^2)}{y^9z^{10}} = \dfrac{120x^8y^4z^3}{y^9z^{10}} = \dfrac{120x^8}{y^5z^7}$

50

3.2   MONOMIALS AND POLYNOMIALS

ALGEBRAIC TERMS

A **monomial** is one algebraic term. A monomial has one sign in front of the expression. For a fractional term, there is no variable in the denominator.

EXAMPLE 1   Each of the following expressions is a monomial: 5 ; x^2 ; $-8y^5$; $5x^3y^4$

In the monomial $5x^3y^4$, the numerical **coefficient** is 5.

EXAMPLE 2   Determine which of the following expressions is a monomial:

(a) $\dfrac{3}{4}x^5$ (b) $\dfrac{3}{4}x^{-5}$

Answers: (a)   The variable is not in the denominator; $\dfrac{3}{4}x^5$ is a monomial.

(b) $\dfrac{3}{4}x^{-5} = \dfrac{3}{4x^5}$, which has x in the denominator. Therefore $\dfrac{3}{4}x^{-5}$ is not a monomial.

A **polynomial** is the sum or difference of two or more monomials. The terms of a polynomial are separated by a positive or a negative sign.

EXAMPLE 3

(a) The polynomial $9x^4 - 5x^2 + 3x - 2$ has four terms.
(b) The polynomial $5x^2 + 3y$ is called a **binomial** (two terms).
(c) The polynomial $3x^2 + 5x - 11$ is called a **trinomial** (three terms).

DEGREE OF A POLYNOMIAL

The **degree of a term** is the sum of the exponents on the variables.

EXAMPLE 4   In the following table, the degree is given for each term.

Term	Degree	Term	Degree
(a) $5x^2$	2nd	(d) $2x^3y^5$	8th
(b) $-8x^2y^3$	5th	(e) $4x$	1st
(c) $-4y^5$	5th	(f) -4	zero

✠ *Helpful Hint* – Observe that a constant, such as -4, has no variable. Therefore the degree for Example 4(f) is zero.

To determine the **degree of a polynomial**, write the degree of each term. The highest of the degrees is the degree of the polynomial.

EXAMPLE 5   Find the degree of the polynomial $2x^5y^3 - 8x^3y^2 + 4y$.

Answer: Degree of $2x^5y^3$: 8th Degree of $-8x^3y^2$: 5th Degree of $4y$: 1st

Therefore $2x^5y^3 - 8x^3y^2 + 4y$ is an 8th degree polynomial.

(3.2)

A monomial is in **standard form** if the variables are in alphabetical order, such as $-2x^3y^2$.
A polynomial is in standard form if (a) each term is in alphabetical order, (b) the terms are listed in decreasing powers for the first variable (alphabetically) of the polynomial and (c) the constant term is last.

EXAMPLE 6 Write each of the following polynomials in standard form:
(a) $7x^2 + 1 + 5x^4 + 3x - 2x^3$ (b) $9y^8 - 5 + 3x^2 + 3xy^4 - 2x^3y^2$

Answers:
(a) Rewrite $7x^2 + 1 + 5x^4 + 3x - 2x^3$ as: $5x^4 - 2x^3 + 7x^2 + 3x + 1$.
(b) Rewrite $9y^8 - 5 + 3x^2 + 3xy^4 - 2x^3y^2$ as: $-2x^3y^2 + 3x^2 + 3xy^4 + 9y^8 - 5$.

ADDITION AND SUBTRACTION OF POLYNOMIALS

Add or subtract the coefficients of the similar terms. Always write the answer in standard form.

EXAMPLE 7 Add $2x^3 + 6x^2 - 7x + 1$ to $5x^3 + 3x^2 - 4x - 7$.

Answer: $(2x^3 + 6x^2 - 7x + 1) + (5x^3 + 3x^2 - 4x - 7) =$

$$2x^3 + 6x^2 - 7x + 1 + 5x^3 + 3x^2 - 4x - 7 = 7x^3 + 9x^2 - 11x - 6$$

EXAMPLE 8 Subtract $2x^3 + 6x^2 - 7x + 1$ from $5x^3 + 3x^2 - 4x - 7$.

Helpful Hint – For subtraction, the "from" polynomial should be written first, followed by a minus symbol, and *the subtracted polynomial must be in parentheses.*

Answer: $(5x^3 + 3x^2 - 4x - 7) - (2x^3 + 6x^2 - 7x + 1) =$

$$5x^3 + 3x^2 - 4x - 7 - 2x^3 - 6x^2 + 7x - 1 = 3x^3 - 3x^2 + 3x - 8$$

MULTIPLICATION OF POLYNOMIALS

Multiplying Monomials

To multiply monomials, refer to Section 3.1, "Property 3 – Basic Rule for Multiplication": Multiply the coefficients and add the exponents for each variable.

EXAMPLE 9 $(5x^3y^2)(2x^3y) = 10x^6y^3$

Distributive Property $A(B + C) = AB + AC$

Multiply the outside term times each inside term. As each pair of terms is multiplied, add the exponents for each variable.

EXAMPLE 10 $5x^3y^2(2x^3y + 5x^2y^2 - 3x + 4)$

$$= (5x^3y^2)(2x^3y) + (5x^3y^2)(5x^2y^2) + (5x^3y^2)(-3x) + (5x^3y^2)(4)$$
$$= 10x^6y^3 + 25x^5y^4 - 15x^4y^2 + 20x^3y^2$$

(Polynomial)(Polynomial)

Multiply each term of the first polynomial times all of the second polynomial. Then combine the similar terms, and write the final answer in standard form.

EXAMPLE 11   Multiply: $(2x - 3)(x^2 + 3x - 4)$

Answer: $(2x - 3)(x^2 + 3x - 4) = 2x(x^2 + 3x - 4) - 3(x^2 + 3x - 4)$

$$= 2x^3 + 6x^2 - 8x - 3x^2 - 9x + 12 = 2x^3 + 3x^2 - 17x + 12$$

(Binomial)(Binomial)

As shown in the example above, multiply each term of the first binomial times all of the second binomial. Combine the similar terms, writing the final answer in standard form.

EXAMPLE 12   Multiply: $(2x + 3)(4x - 5)$

✖ *Helpful Hints* – The product (Binomial)(Binomial) is also called the "**F. O. I. L.**" Method.

Multiply the **F**irst term of each binomial: $(2x)(4x)$
Multiply the **O**utside terms of the binomials: $(2x)(- 5)$
Multiply the **I**nside terms of the binomials: $(3)(4x)$
Multiply the **L**ast term of each binomial: $(3)(- 5)$
Then combine the similar terms.

Answer: $(2x + 3)(4x - 5) = 8x^2 - 10x + 12x - 15 = 8x^2 + 2x - 15$

Observation - - Normally a *(Binomial)(Binomial) = A Trinomial* answer. An important exception, "Multiplying Conjugates", will be presented later in this section.

(Binomial)^2 = A Trinomial

To square a binomial, rewrite the problem as a binomial times itself, (Binomial)(Binomial), and use F.O.I.L. The answer is always a trinomial.

EXAMPLE 13   Multiply: $(8s - 3t)^2$

Answer: $(8s - 3t)^2 = (8s - 3t)(8s - 3t) = 64s^2 - 24st - 24st + 9t^2 = 64s^2 - 48st + 9t^2$

Multiplying Conjugates   (x + y)(x – y) = "Square – Square"

Conjugates are two binomials which have the same terms but different signs, such as $(x + y)(x - y)$. If conjugates are multiplied using the F. O. I. L. Method, the two middle terms cancel out and the result is a square term minus a square term:

$$(x + y)(x - y) = x^2 - xy + xy - y^2 = \underbrace{x^2 - y^2}$$

Square – Square

✖ *Helpful Hint* – "Multiplying Conjugates"- - This is the only binomial product which has the *Shortcut: The square of the 1st term minus the square of the 2nd term.*

(3.2)

EXAMPLE 14 Multiply the conjugates: $(7x - 5)(7x + 5)$

Answer: Using the shortcut *Square – Square*, $(7x - 5)(7x + 5) = 49x^2 - 25$

Other Examples of Multiplying Polynomials

EXAMPLE 15 Multiply: $(x + y)^3$

Helpful Hint – Rewrite the problem as the product of the same three binomials. Multiply the first two binomials using F.O.I.L. Multiply the result times the third binomial and combine similar terms.

Answer:

$$\begin{aligned}(x + y)^3 &= (x + y)(x + y)(x + y) = (x^2 + xy + xy + y^2)(x + y)\\ &= (x^2 + 2xy + y^2)(x + y)\\ &= x^2(x + y) + 2xy(x + y) + y^2(x + y)\\ &= x^3 + x^2y + 2x^2y + 2xy^2 + xy^2 + y^3\\ &= x^3 + 3x^2y + 3xy^2 + y^3\end{aligned}$$

EXAMPLE 16 Multiply: $(2x^2 - x + 3)^2$

Answer:

$$\begin{aligned}(2x^2 - x + 3)^2 &= (2x^2 - x + 3)(2x^2 - x + 3)\\ &= 2x^2(2x^2 - x + 3) - x(2x^2 - x + 3) + 3(2x^2 - x + 3)\\ &= 4x^4 - 2x^3 + 6x^2\\ &\qquad - 2x^3 + x^2 - 3x\\ &\qquad\qquad + 6x^2 - 3x + 9\\ &= 4x^4 - 4x^3 + 13x^2 - 6x + 9\end{aligned}$$

DIVISION OF POLYNOMIALS

A Polynomial Divided By A Monomial

Rewrite the expression as separate fractions: $\frac{A+B}{C} = \frac{A}{C} + \frac{B}{C}$. Then divide each term.

EXAMPLE 17 Divide: $\frac{7x^3y^3 - 21x^2y^3 + 14xy}{7x^3y^3}$

Answer:

$$\frac{7x^3y^3 - 21x^2y^3 + 14xy}{7x^3y^3} = \frac{7x^3y^3}{7x^3y^3} - \frac{21x^2y^3}{7x^3y^3} + \frac{14xy}{7x^3y^3}$$

$$= 1 - \frac{3}{x} + \frac{2}{x^2y^2}$$

A Polynomial Divided By A Polynomial

The division of two polynomials is given as $\dfrac{\text{polynomial}}{\text{polynomial}}$ or (polynomial) ÷ (polynomial).

When the divisor (denominator) is a polynomial, we use the **Long Division Method** to simplify the expression:

$$\text{Divisor}\overline{)\text{Numerator}} \quad = \quad \text{Quotient} + \frac{\text{Remainder}}{\text{Divisor}}$$

In the example below, the steps for the long division procedure are: *Divide, Multiply, Subtract and Bring down.* Repeat the procedure until the problem is finished. Make a fraction of the remainder. (If the remainder is zero, omit the fraction.)

EXAMPLE 18   Divide: $\dfrac{6x^3 + 7x^2 - 4x + 2}{3x + 2}$

Answer: Rewrite the problem for long division.

(a) *Divide* "1st term into 1st term".

    Divide 3x into $6x^3$ to get $2x^2$.
Multiply $2x^2$ times the divisor 3x + 2.

$$\begin{array}{r} 2x^2 \\ 3x+2\overline{)6x^3 + 7x^2 - 4x + 2} \\ \underline{6x^3 + 4x^2} \end{array}$$

(b)

Subtract [change the signs of the last row of step (a)].
Simplify. *Bring down* the next term, – 4x.

$$\begin{array}{r} 2x^2 \\ 3x+2\overline{)6x^3 + 7x^2 - 4x + 2} \\ \underline{-6x^3 - 4x^2} \\ 3x^2 - 4x \end{array}$$

(c) *Divide* 3x into $3x^2$ to get positive x.

Multiply x times the divisor 3x + 2.

$$\begin{array}{r} 2x^2 + x \\ 3x+2\overline{)6x^3 + 7x^2 - 4x + 2} \\ \underline{-6x^3 - 4x^2} \\ 3x^2 - 4x \\ \underline{3x^2 + 2x} \end{array}$$

(d)

Subtract [change the signs of the last row of step (c)].
Simplify. *Bring down* the next term, + 2.

$$\begin{array}{r} 2x^2 + x \\ 3x+2\overline{)6x^3 + 7x^2 - 4x + 2} \\ \underline{-6x^3 - 4x^2} \\ 3x^2 - 4x \\ \underline{-3x^2 - 2x} \\ -6x + 2 \end{array}$$

(Repeat the procedure again to get the following answer lines.)

(3.2)

(e) *Divide* 3x into – 6x to get – 2.

$$
\begin{array}{r}
2x^2 + x - 2 \\
3x+2 \overline{\smash{)}\ 6x^3 + 7x^2 - 4x + 2} \\
\underline{-6x^3 - 4x^2} \\
3x^2 - 4x \\
\underline{-3x^2 - 2x} \\
-6x + 2 \\
\underline{-6x - 4}
\end{array}
$$

Multiply – 2 times the divisor 3x + 2.

(f)

$$
\begin{array}{r}
2x^2 + x - 2 \\
3x+2 \overline{\smash{)}\ 6x^3 + 7x^2 - 4x + 2} \\
\underline{-6x^3 - 4x^2} \\
3x^2 - 4x \\
\underline{-3x^2 - 2x} \\
-6x + 2 \\
\underline{+6x + 4} \\
6
\end{array}
$$

Subtract [change the signs of the last row of step (e)].
Simplify. There are *no terms to bring down.*

(g) The quotient is $2x^2 + x - 2$ and the remainder is 6.

Therefore $\dfrac{6x^3 + 7x^2 - 4x + 2}{3x + 2} = 2x^2 + x - 2 + \dfrac{6}{3x + 2}$.

✠ *Helpful Hint* – To **check long division**, multiply the quotient times the divisor and add the remainder. The result should be the numerator of the original problem. Let's check Example 2:

(Quotient)(Divisor) + Remainder =

$(2x^2 + x - 2)(3x + 2) + 6 = 2x^2(3x + 2) + x(3x + 2) - 2(3x + 2) + 6$

$= 6x^3 + 4x^2 + 3x^2 + 2x - 6x - 4 + 6$

$= 6x^3 + 7x^2 - 4x + 2$ (the numerator of the problem)

Special Case -- For Long Division of Polynomials

If a polynomial in standard form has missing terms, fill in a zero for each missing term.

EXAMPLE 19   Rewrite the polynomial $8x^3 - 27$ for long division.

Answer: The 3rd degree polynomial $8x^3 - 27$ is missing the x^2 term and the x term.
For long division, rewrite $8x^3 - 27$ as: $8x^3 + 0 + 0 - 27$.

EXAMPLE 20   Rewrite the polynomial $2x^5 - x^2 - 7$ for long division.

Answer: The 5th degree polynomial $2x^5 - x^2 - 7$ is missing the x^4 term, the x^3 term and the x term.
For long division, we rewrite $2x^5 - x^2 - 7$ as: $2x^5 + 0 + 0 - x^2 + 0 - 7$.

(3.2)

EXAMPLE 21   Divide: $(8x^3 - 27) \div (2x - 3)$
Answer: Rewrite the problem for long division.

(a) *Divide* "1st term into 1st term".

    Divide 2x into $8x^3$ to get $4x^2$.
Multiply $4x^2$ times the divisor $2x - 3$.

$$\begin{array}{r} 4x^2 \qquad\qquad\qquad \\ 2x-3\overline{)8x^3 \; + \; 0 \; + \; 0 \; - \; 27} \\ \underline{8x^3 \; - \; 12x^2} \qquad\qquad \end{array}$$

(b)

Subtract [change the signs of the last row of step (a)].
Simplify. *Bring down* the next term, 0.

$$\begin{array}{r} 4x^2 \qquad\qquad\qquad \\ 2x-3\overline{)8x^3 \; + \; 0 \; + \; 0 \; - \; 27} \\ \underline{-8x^3 \; + \; 12x^2} \qquad\qquad \\ 12x^2 \; + \; 0 \qquad\qquad \end{array}$$

(c) *Divide* 2x into $12x^2$ to get positive 6x.

Multiply 6x times the divisor $2x - 3$.

$$\begin{array}{r} 4x^2 + 6x \qquad\qquad \\ 2x-3\overline{)8x^3 \; + \; 0 \; + \; 0 \; - \; 27} \\ \underline{-8x^3 \; + \; 12x^2} \qquad\qquad \\ 12x^2 \; + \; 0 \qquad\qquad \\ \underline{12x^2 \; - \; 18x} \qquad\qquad \end{array}$$

(d)

Subtract [change the signs of the last row step (c)].
Simplify. *Bring down* the next term, -27.

$$\begin{array}{r} 4x^2 + 6x \qquad\qquad \\ 2x-3\overline{)8x^3 \; + \; 0 \; + \; 0 \; - \; 27} \\ \underline{-8x^3 \; + \; 12x^2} \qquad\qquad \\ 12x^2 \; + \; 0 \qquad\qquad \\ \underline{-12x^2 \; + \; 18x} \qquad\qquad \\ 18x \; - \; 27 \qquad \end{array}$$

(e) *Divide* 2x into 18x to get positive 9.

Multiply 9 times the divisor $2x - 3$ to get $18x - 27$.

$$\begin{array}{r} 4x^2 + 6x + 9 \\ 2x-3\overline{)8x^3 \; + \; 0 \; + \; 0 \; - \; 27} \\ \underline{-8x^3 \; + \; 12x^2} \qquad\qquad \\ 12x^2 \; + \; 0 \qquad\qquad \\ \underline{-12x^2 \; + \; 18x} \qquad\qquad \\ 18x \; - \; 27 \qquad \\ \underline{18x \; - \; 27} \qquad \end{array}$$

(f)

Subtract [change the signs of the last row of step (e)].
Simplify. There are *no terms to bring down*.
The remainder is 0 (no remainder).

$$\begin{array}{r} 4x^2 + 6x + 9 \\ 2x-3\overline{)8x^3 \; + \; 0 \; + \; 0 \; - \; 27} \\ \underline{-8x^3 \; + \; 12x^2} \qquad\qquad \\ 12x^2 \; + \; 0 \qquad\qquad \\ \underline{-12x^2 \; + \; 18x} \qquad\qquad \\ 18x \; - \; 27 \qquad \\ \underline{-18x \; + \; 27} \qquad \\ 0 \qquad \end{array}$$

(g) Therefore $(8x^3 - 27) \div (2x - 3) = 4x^2 + 6x + 9$.

3.3   FACTORING POLYNOMIALS

Factoring in one of the major skills of College Algebra. "**Factor a polynomial**" means rewrite the given polynomial as a multiplication expression. In this section, we will review six factoring methods. The following examples illustrate the factoring methods.

GREATEST COMMON FACTOR METHOD   "GCF"

The Greatest Common Factor Method is the reverse of the Distributive Property. Given a polynomial, the greatest common factor is the *highest monomial* factor which all of the terms have in common.

> Greatest Common Factor Method: **Polynomial = GCF(Other Factors)**

EXAMPLE 1   Factor: $10x^5y + 20x^4y^2 - 15x^2y^4$

Answer: The GCF for this polynomial is $5x^2y$.

Therefore $10x^5y + 20x^4y^2 - 15x^2y^4 = 5x^2y(2x^3 + 4x^2y - 3y^3)$.

✠ *Helpful Hint* – Observe that the correct answer to the GCF Method has no common factors inside the parentheses.

DIFFERENCE OF SQUARES METHOD   "SQ – SQ"

This factoring method is for a perfect square *minus* another perfect square. The Difference of Squares will factor as the product of two binomial conjugates.

> Difference of Squares Method: **SQ – SQ = (+   )(–   )**
> conjugates

Remember that conjugates are two binomials which have the same terms but different signs. Given the difference of two squares, identify the base of each square term. Then use these results to form the two conjugates. (Either sign may be used first.)

EXAMPLE 2   Factor: $a^2 - b^2$.

Answer: Since $a^2 = (a)^2$, the base for a^2 is a. Similarly, $b^2 = (b)^2$ and the base for b^2 is b.
Therefore $a^2 - b^2 = (a - b)(a + b)$

EXAMPLE 3   Factor: $25m^2 - 16n^2$.

Answer: $25m^2 = (5m)^2$ and $16n^2 = (4n)^2 \rightarrow 25m^2 - 16n^2 = (5m - 4n)(5m + 4n)$

EXAMPLE 4   Factor: $64x^6 - 81y^4$.

Answer: $64x^6 = (8x^3)^2$ and $81y^4 = (9y^2)^2 \rightarrow 64x^6 - 81y^4 = (8x^3 - 9y^2)(8x^3 + 9y^2)$

✠ *Helpful Hint* – **The sum of two squares, "Square + Square" cannot be factored.**

EXAMPLE 5   (a) $a^2 + b^2$ cannot be factored.   (b) $25m^2 + 16n^2$ cannot be factored.

(c) $64x^6 + 81y^4$ cannot be factored.

TRINOMIAL METHOD   "TRI"

For the Trinomial Method rewrite the given trinomial as the product of two binomials.

> **Trinomial Method: TRInomial** = (Binomial)(Binomial)

If a trinomial can be factored, there is only <u>one</u> correct pair of binomial factors.

Type-1 Trinomial: The trinomial has 1 as the leading coefficient. For the binomials, choose the *correct numbers* and use the *correct signs* as shown in the following examples.

EXAMPLE 6   Factor the trinomial $x^2 + 4x + 3$.

Answer: Rewrite the trinomial as the product of two binomials. $x^2 + 4x + 3$

(a) Set up two sets of parentheses for the binomials.   (   )(   )
    Choose two factors of the first term: $x^2 = x \cdot x$ (x   )(x   )

(b) Choose two factors of the last term: $3 = 3 \cdot 1$ or $1 \cdot 3$ (x   3)(x   1)
(c) In the workspace below the parentheses,   Check: 3x inside product
    multiply the outside terms. Multiply the inside terms.   1x outside product

(d) <u>Check</u> the outside and inside products: (x   3)(x   1)
In the trinomial, *if the last sign is positive, add* the outside product   Check: + 3x inside product
and inside product to match the middle term of the polynomial: + 1x outside product
    $+ 3x + 1x = + 4x$
(e) Copy the signs from the workspace into the binomials. $x^2 + 4x + 3 = (x + 3)(x + 1)$

Observe that when the last sign of the trinomial is positive, the binomials have the same signs, either two positive signs or two negative signs.

EXAMPLE 7   Factor the trinomial $x^2 - x - 12$.

Answer: Rewrite the trinomial as the product of two binomials. $x^2 - x - 12$

(a) Set up two sets of parentheses for the binomials.   (   )(   )
    Choose two factors of the first term: $x^2 = x \cdot x$ (x   )(x   )

(b) Choose two factors of the last term: The factors of 12 are
    $1 \cdot 12$, or $2 \cdot 6$, or $3 \cdot 4$. The correct pair of factors is $12 = 3 \cdot 4$.   (x   3)(x   4)
(c) In the workspace below the parentheses,   Check: 3x inside product
    multiply the outside terms. Multiply the inside terms.   4x outside product

(d) <u>Check</u> the outside and inside products: (x   3)(x   4)
In the trinomial, *if the last sign is negative, subtract* the outside   Check: + 3x inside product
product and inside product to match the middle term of the   – 4x outside product
polynomial: $+ 3x - 4x = - x$

(e) Copy the signs from the workspace into the binomials. $x^2 - x - 12 = (x + 3)(x - 4)$

Observe that when the last sign of the trinomial is negative, the binomials have the different signs. Make sure the correct sign is in each binomial.

(3.3)

Type-2 Trinomial: The trinomial has a leading coefficient other than 1. For the binomial factors, choose the *correct numbers*, in the correct order, and use the *correct signs*.

EXAMPLE 8 Factor the trinomial $2x^2 - 5x + 3$.

Helpful Hint – For Type-2 Trinomials: If the checking step for the outside and inside products does not match the middle term of the polynomial, switch the factors of the last term. Then check the problem again.

Answer: Rewrite the trinomial as the product of two binomials. $2x^2 - 5x + 3$

(a) Set up two sets of parentheses for the binomials. $(\quad)(\quad)$
Choose two factors of the first term: $2x^2 = 2x \cdot x$ $(2x \quad)(x \quad)$

(b) Choose two factors of the last term: $3 = 1 \cdot 3$ $(2x \quad 1)(x \quad 3)$
(c) The last sign of the trinomial is positive.
Check the outside and inside products by adding:
$-1x - 6x = -7x$ which does not match the middle term

Check: $-1x$ inside product
$-6x$ outside product
(This does not check.)

(d) Try again: Switch the factors of the last term: $3 = 3 \cdot 1$ $(2x \quad \mathbf{3})(x \quad \mathbf{1})$
Check the outside and inside products again by adding:
$-3x - 2x = -5x$ which matches the middle term

Check: $-3x$ inside product
$-2x$ outside product

(e) Copy the signs from the workspace into the binomials. $2x^2 - 5x + 3 = (2x - 3)(x - 1)$

EXAMPLE 9 Factor the trinomial $6x^2 - 11x - 10$.

Answer: Rewrite the trinomial as the product of two binomials. $6x^2 - 11x - 10$

(a) Set up two sets of parentheses for the binomials. $(\quad)(\quad)$
Choose two factors of the first term: $6x^2 = 6x \cdot x$ or $6x^2 = 2x \cdot 3x$.
Try the more popular product $2x \cdot 3x$. $(2x \quad)(3x \quad)$

(b) Choose two factors of the last term: $10 = 1 \cdot 10$ or $10 = 2 \cdot 5$
Try the more popular product $2 \cdot 5$. $(2x \quad 2)(3x \quad 5)$
(c) The last sign of the trinomial is negative.
Check the outside and inside products by subtracting:
$+6x - 10x = -4x$ which does not match the middle term

Check: $+6x$ inside product
$-10x$ outside product
(This does not check.)

(d) Try again: Switch the factors of the last term: $10 = 5 \cdot 2$ $(2x \quad \mathbf{5})(3x \quad \mathbf{2})$
Check the outside and inside products again by subtracting:
$-15x + 4x = -11x$ which matches the middle term

Check: $-15x$ inside product
$+4x$ outside product

(e) Copy the signs from the workspace into the binomials. $6x^2 - 11x - 10 = (2x - 5)(3x + 2)$

Helpful Hint - If you switch the factors of the last term and the checking step for the outside and inside products does not match the middle term of the polynomial, change the factors for the first term, or, change the factors for the last term. Then check the problem again.

GROUPING METHOD   "GR"

To use the Grouping Method, the polynomial must have at least 4 terms. Factor by Grouping also requires two answer lines. The final result is the product of two polynomials. Each polynomial must be in parentheses.

> Grouping Method: **Four or more terms** = (Polynomial)(Polynomial)

EXAMPLE 10   Factor: $a^2 - 3ab + 5a - 15b$

Answer: Observe that there is no greatest common factor for all four terms. Factor the polynomial *in pairs,* two terms at a time.

$$\underline{a^2 - 3ab} + \underline{5a - 15b}$$

(a) Factor the first two terms.   1st answer line: $= a\underline{(a - 3b)} + 5\underline{(a - 3b)}$
Bring down the *plus sign* in the middle.
Factor the second pair of terms.

(b) Both parts of the first answer line have a
binomial common factor in parentheses, $(a - 3b)$.
Factor out the binomial common factor.   2nd answer line: $= (a - 3b)(a + 5)$
Write the remaining terms, a and $+ 5$,
in parentheses as the second factor.

EXAMPLE 11   Factor: $xy + 4y^2 - 2x - 8y$

Answer: Factor the polynomial in pairs, two terms at a time.

$$\underline{xy + 4y^2} - \underline{2x - 8y}$$

(a) Factor the first two terms.   1st answer line: $= y\underline{(x + 4y)} - 2\underline{(x + 4y)}$
Bring down the *minus sign* in the middle.
Factor the second pair of terms.

(b) Both parts of the first answer line have a
binomial common factor in parentheses, $(x + 4y)$.
Factor out the binomial common factor.   2nd answer line: $= (x + 4y)(y - 2)$
Write the remaining terms, y and $- 2$,
in parentheses as the second factor.

EXAMPLE 12   Factor by Grouping: $ak - ap - bk + bp + ck - cp$

Answer: Observe that the polynomial has six terms. Factor the polynomial in pairs, two terms at a time.

$$\underline{ak - ap} - \underline{bk + bp} + \underline{ck - cp}$$

$$= a\underline{(k - p)} - b\underline{(k - p)} + c\underline{(k - p)}$$

$$= (k - p)(a - b + c)$$

(3.3)

SUM OF CUBES METHOD "CU + CU"

✠ *Helpful Hints* – **Review of Cubes** (1) The first five perfect cubes are 1, 8, 27, 64, and 125. (2) Observe that the cube of a positive number is positive, and the cube of a negative number is negative: $(2)^3 = 8$, $(-2)^3 = -8$; $(3)^3 = 27$, $(-3)^3 = -27$. (3) For variables, the exponent of a cube is a multiple of 3: $x^3, x^6, x^9, x^{12}, x^{15}, \ldots$. In other words, $x^6 = (x^2)^3$, $x^9 = (x^3)^3$, $x^{12} = (x^4)^3$, $x^{15} = (x^5)^3$.

The Sum of Cubes Method is for a perfect cube plus another perfect cube. "Cube + Cube" will factor as a binomial times a trinomial:

$$\text{Sum of Cubes Method: } \textbf{CU + CU = (Binomial)(Trinomial)}$$

Important: *For Cube + Cube, the trinomial cannot be factored.*

EXAMPLE 13 Factor: $x^3 + y^3$

Steps:

(a) To get the binomial factor, identify the base of each cubic term. $\quad x^3 + y^3 = (x + y)(\quad ? \quad)$
The base for x^3 is x; the base for y^3 is y.

(b) Use the *binomial* to set up the trinomial factor. $\quad = (x + y)(x^2 \;\; -xy \;\; + \;\; y^2)$
"SQ" - - Square the first term of the binomial: $(x)^2 = x^2$ $\qquad\qquad$ SQ $\quad$ Mult $\quad$ SQ
"Mult & Chg" - - Multiply both terms and change the sign: $\qquad\qquad\qquad$ & Chg
$\qquad (x)(y) = xy$; change the sign $\rightarrow -xy$
"SQ" - - Square the second term of the binomial: $(y)^2 = y^2$

Answer: $x^3 + y^3 = (x + y)(x^2 - xy + y^2)$

EXAMPLE 14 Factor: $a^3 + 125$

(a) To get the binomial factor, identify the base of each cubic term. $\quad a^3 + 125 = (a + 5)(\quad ? \quad)$
The base for a^3 is a. Since $125 = (5)^3$, the base for 125 is 5.

(b) Use the *binomial* to set up the trinomial factor. $\quad = (a + 5)(a^2 \;\; -5a \;\; + \;\; 25)$
"SQ" - - Square the first term of the binomial: $(a)^2 = a^2$ $\qquad\qquad$ SQ $\quad$ Mult $\quad$ SQ
"Mult & Chg" - - Multiply both terms and change the sign: $\qquad\qquad\qquad$ & Chg
$\qquad (a)(5) = 5a$; change the sign $\rightarrow -5a$
"SQ" - - Square the second term of the binomial: $(5)^2 = 25$

Answer: $a^3 + 125 = (a + 5)(a^2 - 5a + 25)$

EXAMPLE 15 Factor: $8r^3 + s^3$

(a) To get the binomial factor, identify the base of each cubic term. $\quad 8r^3 + s^3 = (2r + s)(\quad ? \quad)$
Since $8r^3 = (2r)^3$, the base for $8r^3$ is 2r. The base for s^3 is s.

(b) Use the *binomial* to set up the trinomial factor. $\quad = (2r + s)(4r^2 \;\; -2rs \;\; + \;\; s^2)$
"SQ" - - Square the first term of the binomial: $(2r)^2 = 4r^2$ $\qquad\qquad$ SQ $\quad$ Mult $\quad$ SQ
"Mult & Chg" - - Multiply both terms and change the sign: $\qquad\qquad\qquad$ & Chg
$\qquad (2r)(s) = 2rs$; change the sign $\rightarrow -2rs$
"SQ" - - Square the second term of the binomial: $(s)^2 = s^2$

Answer: $8r^3 + s^3 = (2r + s)(4r^2 - 2rs + s^2)$

(3.3)

DIFFERENCE OF CUBES METHOD   "CU – CU"

The Difference of Cubes Method is for a perfect cube minus another perfect cube. "Cube – Cube" will factor as a binomial times a trinomial:

Difference of Cubes Method: **CU – CU = (Binomial)(Trinomial)**

Important: *For Cube – Cube, the trinomial cannot be factored.*

Use the same steps as shown in Examples 13-15.

EXAMPLE 16   Factor: $x^3 - y^3$

Steps:
(a) To get the binomial factor, identify the base of each cubic term. $x^3 - y^3 = (x - y)(\quad ? \quad)$
    The base for x^3 is x. Since $- y^3 = (- y)^3$, the base for $- y^3$ is $- y$.

(b) Use the *binomial* to set up the trinomial factor. $= (x - y)(x^2 + xy + y^2)$
    "SQ" - - Square the first term of the binomial: $(x)^2 = x^2$ SQ Mult SQ
    "Mult & Chg" - - Multiply both terms and change the sign: & Chg
                $(x)(- y) = - xy$; change the sign $\rightarrow$ xy
    "SQ" - - Square the second term of the binomial: $(- y)^2 = + y^2$

Answer: $x^3 - y^3 = (x - y)(x^2 + xy + y^2)$

EXAMPLE 17   Factor: $z^3 - 1$

(a) To get the binomial factor, identify the base of each cubic term. $z^3 - 1 = (z - 1)(\quad ? \quad)$
    The base for z^3 is z. Since $- 1 = (- 1)^3$, the base for $- 1$ is $- 1$.

(b) Use the *binomial* to set up the trinomial factor. $= (z - 1)(z^2 + z + 1)$
    "SQ" - - Square the first term of the binomial: $(z)^2 = z^2$ SQ Mult SQ
    "Mult & Chg" - - Multiply both terms and change the sign.   & Chg
                $(z)(- 1) = - z$; change the sign $\rightarrow$ z
    "SQ" - - Square the second term of the binomial: $(- 1)^2 = + 1$

Answer: $z^3 - 1 = (z - 1)(z^2 + z + 1)$

EXAMPLE 18   Factor: $27a^3 - b^3$

(a) To get the binomial factor, identify the base of each cubic term. $27a^3 - b^3 = (3a - b)(\quad ? \quad)$
    Since $27a^3 = (3a)^3$, the base for $27a^3$ is 3a; the base for $- b^3$ is $- b$.

(b) Use the *binomial* to set up the trinomial factor. $= (3a - b)(9a^2 + 3ab + b^2)$
    "SQ" - - Square the first term of the binomial: $(3a)^2 = 9a^2$ SQ Mult SQ
    "Mult & Chg" - - Multiply both terms and change the sign.   & Chg
                $(3a)(- b) = - 3ab$; change the sign $\rightarrow$ 3ab
    "SQ" - - Square the second term of the binomial: $(- b)^2 = + b^2$

Answer: $27a^3 - b^3 = (3a - b)(9a^2 + 3ab + b^2)$

(3.3)

USING MORE THAN ONE FACTORING METHOD

To factor a polynomial completely, use every factoring method which applies to the polynomial. The key is to recognize the correct factoring method and write the factored answer in the correct format.

Factoring Strategy:

- Do **GCF** first.
- Then look at the size of the polynomial:

2 terms	$\Rightarrow$	Try **SQ – SQ**.	(If there are cubes, try **CU + CU** or **CU – CU**.)
3 terms	$\Rightarrow$	Try **TRI**nomial	
4 or more terms	$\Rightarrow$	Try **GR**ouping	

- If none of the methods apply, write "**prime**" or "cannot be factored."

For the following problems, the factoring method to be used will be stated at the left and the answer line will be given at the right.

EXAMPLE 19   Factor completely: $2x^3 - 12x^2 - 80x$

Answer: Start with the Greatest Common Factor Method. Then for 3 terms in parentheses, use the Trinomial Method.

GCF $2x^3 - 12x^2 - 80x$ Factor out the Greatest Common Factor 2x.

TRI $= 2x(\underline{x^2 - 6x - 40})$ Factor the Trinomial in the parentheses.

           $= 2x(x - 10)(x + 4)$ Bring down *all* of the factors.

EXAMPLE 20   Factor completely: $16x^5 - 81xy^4$

Answer: Start with the Greatest Common Factor Method. Then for 2 terms in parentheses, consider using the Difference of Squares Method.

GCF $16x^5 - 81xy^4$ Factor out the Greatest Common Factor x.

SQ – SQ $= x(\underline{16x^4 - 81y^4})$ Factor the SQ – SQ in the parentheses.

SQ – SQ again $= x\,(\underline{4x^2 - 9y^2})(4x^2 + 9y^2)$ Factor another SQ – SQ in the parentheses.

                 $= x(2x - 3y)(2x + 3y)(4x^2 + 9y^2)$ Bring down all of the factors.

EXAMPLE 21   Factor completely: $x^2y^2 + x^2 - 4y^2 - 4$

Answer: The polynomial has 4 terms. Start with the Grouping Method (two answer lines). Then for 2 terms in parentheses, consider using the Difference of Squares Method.

GR $\underline{x^2y^2 + x^2} - \underline{4y^2 - 4}$ Factor each pair of terms.

(2 lines) $= x^2(y^2 + 1) - 4\,(y^2 + 1)$ Factor out the binomial gcf $(y^2 + 1)$.

SQ – SQ $= (y^2 + 1)(\underline{x^2 - 4})$ Factor the SQ – SQ in the parentheses.

           $= (y^2 + 1)(x - 2)(x + 2)$ Bring down all of the factors.

(3.4)

EXAMPLE 22   Factor completely: $x^2 + 3x + 3$

Answer: The polynomial is a trinomial (3 terms). Consider using the Trinomial Method.

TRI $x^2 + 3x + 3$

(x   3)(x   1)
    + 3x   inside product
    + 1x   outside product

First, the only factors of 3 are $3 \cdot 1$ or $1 \cdot 3$. Next, the last sign of the trinomial is positive. When we check the outside and inside products by adding, $+ 1x + 3x = 4x$. This does not match the middle term of the trinomial, which is 3x.

The trinomial $x^2 + 3x + 3$ is prime.

EXAMPLE 23   Factor completely: $x^5 - 17x^3y^2 + 16xy^4$

Answer: Start with the Greatest Common Factor Method. Then for 3 terms in parentheses, use the Trinomial Method. Finally, for 2 terms in parentheses consider using the Difference of Squares Method.

GCF $x^5 - 17x^3y^2 + 16xy^4$ Factor out the Greatest Common Factor x.

TRI $= x(x^4 - 17x^2y^2 + 16y^4)$ Factor the Trinomial in the parentheses.

SQ – SQ twice $= x(x^2 - 16y^2)(x^2 - y^2)$ Factor each SQ – SQ in the parentheses.

$= x(x - 4y)(x + 4y)(x - y)(x + y)$ Bring down all of the factors.

3.4   POLYNOMIAL EQUATIONS (FACTORING METHOD)

In Chapter 1 we studied the **linear equation**, $ax + b = c$, which is a first-degree equation. A linear equation has no exponents and it usually has one root (solution).

EXAMPLE 1   The linear equation $3x + 5 = 0$ has one root:

$$3x + 5 = 0$$
$$\underline{-5 \quad -5}$$
$$3x = -5$$

$$\frac{3x}{3} = \frac{-5}{3}$$

$$x = -\frac{5}{3}$$

EQUATIONS WITH EXPONENTS

A **quadratic equation**, $ax^2 + bx + c = 0$, is a second-degree equation. It has exactly 2 roots or solutions. A **cubic equation**, $ax^3 + bx^2 + cx + d = 0$, is a third-degree equation. It has exactly 3 roots or solutions.

RULE FOR POLYOMIAL EQUATIONS: The degree of the polynomial equation indicates the exact number of roots or solutions.

(3.4)

SOLVING POLYNOMIAL EQUATIONS BY FACTORING

> Steps:
> * The polynomial *equation must equal 0*.
> * *Factor* the polynomial *completely*.
> * *Set each factor* which contains a variable *equal to 0*.
> * *Solve* each of the smaller linear equations *for **x***.

EXAMPLE 2   Solve the equation: $x^2 - 7x + 12 = 0$.

The equation already equals zero.	$x^2 - 7x + 12 = 0$
Factor completely (using the Trinomial Method).	$(x - 3)(x - 4) = 0$
Set each factor which has a variable equal to 0.	$x - 3 = 0 \qquad x - 4 = 0$
Solve for x.	Answers: $x = 3$ or $x = 4$

EXAMPLE 3   Solve the equation: $6x^2 + 3 = 3 - 10x$.

�֍ ***Helpful Hints*** – (1) This polynomial equation is not equal to zero. Cancel out one side of the equation. Zero may be on the right side or the left side of the equation. (2) When you rewrite a polynomial equation, write the terms in standard form—that is, write the polynomial in order, starting with the highest degree term.

The equation must equal 0.	$6x^2 + 3 \qquad = 3 - 10x$
Cancel out the right side of the equation.	$\qquad\quad -3 + 10x \quad -3 + 10x$
Rewrite the equation in standard form.	$6x^2 + 10x = 0$
Factor completely (using the GCF Method).	$2x(3x + 5) = 0$
Set each factor which has a variable equal to 0.	$2x = 0 \qquad\qquad 3x + 5 = 0$
Solve for *x*.	$\dfrac{2x}{2} = 0 \qquad\qquad 3x = -5$

$$\text{Answers:} \quad x = 0 \quad \text{or} \quad x = -\frac{5}{3}$$

EXAMPLE 4   Solve the equation: $x + 10 = x(x + 4)$

✖ ***Helpful Hints*** – (1) Observe that ***this equation has parentheses but it is not equal to zero***. Therefore, multiply the right side of the equation and rewrite the equation equal to zero. (2) When you rewrite a polynomial equation, the highest degree term should be *positive*. It will be easier to solve the equation. Zero may be on the right side or the left side of the equation.

The equation must equal 0.	$x + 10 = x(x + 4)$
Multiply the right side of the equation.	$x + 10 = x^2 + 4x$
The leading term x^2 is positive.	$-x - 10 \qquad -x - 10$
Cancel the left side of the equation.	$0 = x^2 + 3x - 10$
Factor completely (using the Trinomial Method).	$0 = (x + 5)(x - 2)$
Set each factor which has a variable equal to 0.	$x + 5 = 0 \qquad x - 2 = 0$
Solve for *x*.	Answers: $x = -5$ or $x = 2$

EXAMPLE 5 Solve the equation: $\frac{12}{5}x^3 = 3x^2 - \frac{9}{10}x$

The equation must equal 0.

$$\frac{12}{5}x^3 = 3x^2 - \frac{9}{10}x$$

Eliminate the fractions by multiplying each term times the lowest common denominator, 10.

$$\overset{(2)}{\cancel{(10)}}\frac{12}{\cancel{5}}x^3 = (10)3x^2 - \cancel{(10)}\frac{9}{\cancel{10}}x$$

The leading term $24x^3$ is positive. Cancel out the right side of the equation.

$$24x^3 = 30x^2 - 9x$$
$$-30x^2 - 9x \quad -30x^2 - 9x$$
$$24x^3 - 30x^2 + 9x = 0$$

Factor the polynomial using the GCF Method. $3x\left(8x^2 - 10x + 3\right) = 0$

Factor completely (using the Trinomial Method). $3x(4x - 3)(2x - 1) = 0$

Set each factor which has a variable equal to 0. $3x = 0 \quad 4x - 3 = 0 \quad 2x - 1 = 0$

Solve for each *x*. $\frac{3x}{3} = 0 \quad 4x = 3 \quad 2x = 1$

Answers: $x = 0$, $x = \frac{3}{4}$, $x = \frac{1}{2}$

EXAMPLE 6 Show that $x^5 - 5x^3 + 4x = 0$ has 5 roots.

The equation already equals zero. $x^5 - 5x^3 + 4x = 0$

Factor the polynomial using the GCF Method. $x(x^4 - 5x^2 + 4) = 0$

Factor the polynomial using Trinomial Method. $x(x^2 - 1)(x^2 - 4) = 0$

Factor completely (using the SQ – SQ Method twice). $x(x - 1)(x + 1)(x - 4)(x + 4) = 0$

Set each factor equal to 0. $x = 0 \quad x - 1 = 0 \quad x + 1 = 0 \quad x - 4 = 0 \quad x + 4 = 0$

Solve for *x*. Answers: $x = 0$, $x = 1$, $x = -1$, $x = 4$, $x = -4$

3.5 APPLICATIONS FOR POLYNOMIAL EQUATIONS

For each word problem, write a polynomial equation in one variable based on the words of the problem. Use the appropriate factoring method to solve the equation. Find the answers to the word problem.

Helpful Hints – (1) Geometry involves measurement. When solving geometry word problems, *measurement answers must be positive.* (2) When solving projectile motion problems, *time is positive.*

GEOMETRY – AREA PROBLEMS

For any two-dimensional geometric figure, the **area** is space inside the figure.

Area of a Rectangle

The area of a rectangle is equal to the length times the width.

Formula: **A = LW**

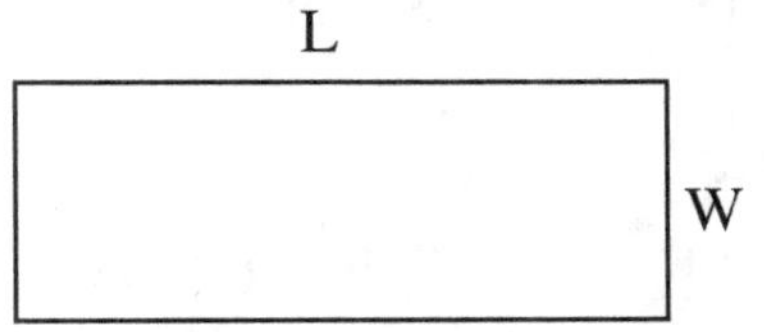

(3.5)

EXAMPLE 1

The length of a rectangle is 1 cm less than 3 times the width. The area of the rectangle is 24 cm^2. Find the length and width of the rectangle.

Table	(Set-up)	(Explanation)
width	x	The length is compared to the width. Identify the width as x.
length	3x – 1	"1 cm less than 3 times the width"
Area	24 cm^2	(Given in the problem)

Formula	A = LW
Equation	24 = (3x – 1)x

The equation must equal 0.	$24 = (3x - 1)x$
Multiply the right side of the equation.	$24 = 3x^2 - x$
The leading term 3x^2 is positive.	$-24 \qquad\qquad -24$
Cancel the left side of the equation.	$0 = 3x^2 - x - 24$

Factor completely (using the Trinomial Method). $0 = (x - 3)(3x + 8)$

Set each factor equal to 0.

Solve for *x*.

$x - 3 = 0$	$3x + 8 = 0$
$x = 3$	$3x = -8$
	$x = -\dfrac{8}{3}$

Answers The width: x = 3 cm. The length: 3x – 1 = 3(3) – 1 = 8 cm	Geometry: The width x must be positive. Delete this result.

Area of a Triangle

In a triangle, the **base** is the bottom side. For the triangle below, the base is side b and the height is h. The area of a triangle is equal to one half times the base, times the height:

Formula: $A = \dfrac{1}{2}\,\mathbf{bh}$

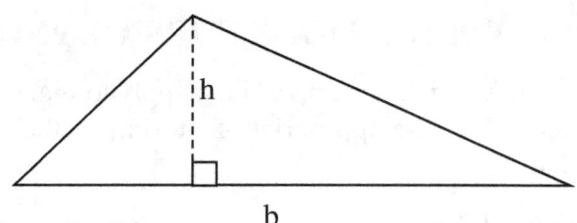

EXAMPLE 2

The height of a triangular wall is 7 feet more than the base. Find the base and the height if the area of the triangular wall is 85 square feet.

Table	(Set-up)	(Explanation)
base	x	The height is compared to the base. Identify the base as x.
height	x + 7	"7 feet more than the base"
Area	85 ft^2	(Given in the problem)

Formula: $A = \dfrac{1}{2}bh$ Equation: $\dfrac{1}{2}x(x + 7) = 85$

The equation must equal 0. $\dfrac{1}{2}x(x + 7) = 85$

Multiply the right side of the equation. $\dfrac{1}{2}x^2 + \dfrac{7}{2}x = 85$

Eliminate the fractions by multiplying each

term times the lowest common denominator, 2. $(2)\dfrac{1}{2}x^2 + (2)\dfrac{7}{2}x = (2)85$

The leading term x^2 is positive. Cancel out the $x^2 + 7x = 170$

right side of the equation. $x^2 + 7x - 170 = 0$

Factor completely (using Trinomial Factoring). $(x - 10)(x + 17) = 0$

Set each factor equal to 0. $x - 10 = 0$ | $x + 17 = 0$

Solve for x. $x = 10$ | $x = -\cancel{17}$

Answers   The base: $x = 10$ feet. | Geometry: The base x must
         The height: $x + 7 = 10 + 7 = 17$ feet | be positive. Delete this result.

GEOMETRY – THE SIDES OF A RIGHT TRIANGLE

In the right triangle below, the **legs** are sides A and B. The **hypotenuse** *must be labeled as side C.* The hypotenuse is across from the right angle; it is also the longest side of the right triangle. The formula used to find the sides of a right triangle is called the **Pythagorean Theorem:**

Formula $\mathbf{A^2 + B^2 = C^2}$

A

C

B

EXAMPLE 3   If leg B = 9 cm and the hypotenuse C = 15 cm, find the length of leg A.

Substitute the given values into the formula. $A^2 + (9)^2 = (15)^2$

Simplify the equation. $A^2 + 81 = 225$
The equation must equal zero. Cancel out the right side of $-225 \quad -225$
the equation. Rewrite the equation in standard form. $A^2 - 144 = 0$

Factor the polynomial (using SQ – SQ Factoring). $(A - 12)(A + 12) = 0$

Set each factor equal to 0. $A - 12 = 0$ | $A + 12 = 0$
Solve for x. $A = 12$ | $A = -\cancel{12}$

Answer   Leg A = 12 cm. | Geometry: Leg A must be
                                       | positive. Delete this result.

(3.5)

EXAMPLE 4   Find the length of each side of the right triangle below (given in millimeters).

✖ *__Helpful Hint__* – The side labeled "x + 1" is across from the right angle. *This side is the hypotenuse and must be labeled as side C.*

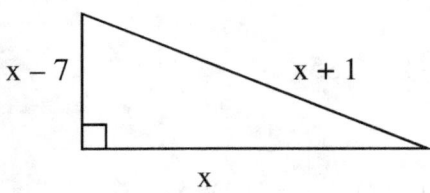

Leg A is the side *labeled x – 7*.

Leg B is the side labeled *x*.

Hypotenuse C is the side labeled *x + 1*.

Substitute the given expressions into the formula.

Simplify the equation using the F.O.I.L. Method.

Combine the similar terms on the left side.
The equation must equal zero. Cancel out the right side of the equation. Rewrite the equation in standard form.

Factor the polynomial (using the Trinomial Method).

Formula: $A^2 + B^2 = C^2$

$(x-7)^2 + x^2 = (x+1)^2$

$x^2 + (x-7)(x-7) = (x+1)(x+1)$

$x^2 + x^2 - 14x + 49 = x^2 + 2x + 1$

$2x^2 - 14x + 49 = x^2 + 2x + 1$
$-x^2 - 2x - 1 \quad -x^2 - 2x - 1$
$x^2 - 16x + 48 = 0$

$(x-12)(x-4) = 0$

Set each factor equal to 0.

$x - 12 = 0$ | $x - 4 = 0$

Solve for *x*.

$x = 12$ | $x = 4$

Find the length of the other two sides.

$x - 7 = 5$ | $x - 7 = -3$

$x + 1 = 13$ |

Answers:
The sides of the right triangle are 5 mm, 12 mm and 13 mm.

These answers are false because the sides of a geometric figure must be positive. Delete these results.

PHYSICS: PROJECTILE MOTION

"An object is thrown upward . . ."
Substitute the given number in place of the appropriate variable. Solve the resulting equation.

EXAMPLE 5   From the ground, an object is propelled upward into the air. The height *h* of the object (in feet) after *t* seconds is given by the formula $h = 48t - 16t^2$. (a) Find the time it takes for the object to be 32 feet above the ground. (b) Find the time(s) at which the object is on the ground.

(a) The height *h* = 32 feet.

Substitute the value for height *h* into the formula.

$32 = 48t - 16t^2$

Since the leading term $-16t^2$ is negative, cancel out the right side of the equation.

$-48t + 16t^2 \quad -48t + 16t^2$
$16t^2 - 48t + 32 = 0$

Factor the polynomial using the GCF Method.

$16(t^2 - 3t + 2) = 0$

Factor the polynomial (using the Trinomial Method).

$16(t-1)(t-2) = 0$

70

Set each factor which has a variable equal to 0. $t - 1 = 0$ | $t - 2 = 0$
Solve for t. $t = 1$ | $t = 2$

Answers:
When the object is moving up in the air, it will be 32 feet above the ground at time $t = 1$ second.
When the object is coming down, it will be 32 feet above the ground at time $t = 2$ seconds.

(b) To find the time(s) at which the object is on the ground:

�֎ *Helpful Hint* – The phrase "on the ground" means the height $h = 0$.

Substitute the value for height h into the formula. $0 \ = \ 48t - 16t^2$

Since the leading term $- 16t^2$ is negative, cancel out
  the right side of the equation.
$$- 48t + 16t^2 \qquad - 48t + 16t^2$$
$$16t^2 - 48t \ = \ 0$$

Factor the polynomial using the Greatest Common Factor. $16t(t - 3) \ = \ 0$

Set each factor which has a variable equal to 0. $16t = 0$ | $t - 3 = 0$

Solve for t. $\dfrac{16t}{16} = \dfrac{0}{16}$ | $+3 \quad +3$

                                                                                          $t = 0$ | $t = 3$

Answers:
At the beginning of the problem, the object is on the ground at time $t = 0$ seconds.
When the object is comes down, it hits the ground at time $t = 3$ seconds.

CONSECUTIVE INTEGERS
For the set-up, you may refer to Section 1.3, Example 4.

EXAMPLE 6   The product of two consecutive even integers is 120. Find the numbers.

Table	(Set-up)	(Explanation)
1st integer	x	Always denote the first integer as x.
2nd integer	x + 2	Add 2 to the previous number.

The equation must equal 0. $x(x + 2) \ = \ 120$
  Multiply the right side of the equation. $x^2 + 2x \ = \ 120$
  The leading term x^2 is positive. $- 120 \quad - 120$
  Cancel the right side of the equation. $x^2 + 2x - 120 = 0$

Factor completely (using Trinomial Factoring). $(x + 12)(x - 10) = 0$

Set each factor which has a variable equal to 0. $x + 12 = 0$ | $x - 10 = 0$
Solve for each x. $x = -12$ | $x = 10$

Both -12 and 10 are answers for x. Now find
  answers for $x + 2$. $x + 2 =$ | $x + 2 =$
                                                                                  $-12 + 2 = -10$ | $10 + 2 = 12$

Answers   This problem has *two solution sets*: $\{-12, -10\}$ and $\{10, 12\}$.

71

Chapter 4   Rational Expressions and Equations

4.1   REDUCING FRACTIONS; MULTIPLICATION AND DIVISION OF FRACTIONS

A **rational expression** is a fraction in which the numerator and denominator are monomials or polynomials, and, *the denominator is not equal to zero*. Examples of rational expressions are:

$$\frac{3}{5}, \qquad \frac{2}{x-7}, \qquad \frac{x^3+5}{x^2-2}, \qquad \text{and} \qquad \frac{5x+1}{x^2+3x-2}$$

IF A FRACTION HAS A VARIABLE IN THE DENOMINATOR

A **restricted value** is the number substituted for the variable which causes the denominator to equal zero. To find the restricted value(s) for the denominator, set the denominator not equal to zero and solve for the variable.

EXAMPLE 1   Determine the restricted value(s) for each fraction:

(a) $\dfrac{1}{x}$ (b) $\dfrac{x}{x-3}$ (c) $\dfrac{5x}{x^2+2x-3}$ (d) $\dfrac{x}{x^2+1}$ (e) $\dfrac{3}{5}$

Answers: In a rational expression, the denominator is not equal to zero.

(a) For $\dfrac{1}{x}$, the restricted value is $x \neq 0$.

(b) For $\dfrac{x}{x-3}$, the denominator $x - 3 \neq 0$. Solve for x to get the restricted value $x \neq 3$.

(c) For $\dfrac{5x}{x^2+2x-3}$, factor the trinomial in the denominator to get $\dfrac{5x}{(x+3)(x-2)}$.

     Set each binomial factor not equal to zero and solve for x: $x + 3 \neq 0$ $x - 2 \neq 0$

                                             Restricted values: $x \neq -3$ and $x \neq 2$.

(d) For $\dfrac{x}{x^2+1}$, the denominator cannot be factored. In the denominator, let x represent a real number.

     When a real number is squared, the result is a positive number or 0.

         This means: (Any positive real number)^2 + $1 \neq 0$
                              (Any negative real number)^2 + $1 \neq 0$
                                      Also 0^2 + $1 \neq 0$

     Therefore, for any real number x, the denominator $x^2 + 1 \neq 0$ and there are *no restricted values* for this fraction.

(e) The fraction $\dfrac{3}{5}$ is a numerical fraction. There is no variable in the denominator and the denominator $5 \neq 0$. There are *no restricted values* for a numerical fraction.

REDUCING FRACTIONS

Monomials in the Fraction: Divide the coefficients and subtract the exponents for each variable.

EXAMPLE 2   Reduce the fraction: $\dfrac{12x^2y}{16xy^5}$

Answer: $\dfrac{12x^2y}{16xy^5} = \dfrac{\overset{3}{\cancel{12}}\,\overset{x}{\cancel{x^2}}\,\cancel{y}}{\underset{4}{\cancel{16}}\,\cancel{x}\,\underset{y^4}{\cancel{y^5}}} = \dfrac{3x}{4y^4}$

Polynomials in the Fraction: Factor the numerator and the denominator completely and cancel the common factors. To cancel, one factor must be in the numerator and the other factor must be in the denominator. Also, a monomial factor may cancel another monomial factor. A binomial factor in the numerator may cancel out *the same binomial factor* in the denominator. After the cancellation step, write the remaining factors as the final answer in a separate step.

EXAMPLE 3   Reduce the fraction: $\dfrac{12x+12y}{16x+16y}$

$$\qquad\qquad\text{Step (a)}\qquad\qquad\text{Step (b)}\qquad\qquad\text{Step (c)}$$

Answer: $\dfrac{12x+12y}{16x+16y} = \dfrac{12\,(x+y)}{16\,(x+y)} = \dfrac{\overset{3}{\cancel{12}}\,\cancel{(x+y)}}{\underset{4}{\cancel{16}}\,\cancel{(x+y)}} = \dfrac{3}{4}$

(a) *Factor* the numerator and the denominator using the Greatest Common Factor Method.
(b) *Cancel out* the monomial factors 12 and 16 (dividing by 4). Cancel out the binomial factors (x + y).
(c) Write the result as a separate step.

EXAMPLE 4   Reduce each fraction, if possible:

(a) $\dfrac{2(x+3)}{x(x+3)}$ (b) $\dfrac{2x+3}{2(x+3)}$ (c) $\dfrac{x}{x+3}$ (d) $\dfrac{x+2}{x^2+x+2}$

Answers:

(a) $\dfrac{2(x+3)}{x(x+3)} = \dfrac{2\cancel{(x+3)}}{x\cancel{(x+3)}} = \dfrac{2}{x}$

The monomials 2 and x cannot be canceled. Cancel out the binomial (x + 3).

(b) In the fraction $\dfrac{2x+3}{2(x+3)}$, 2x + 3 is a binomial and 2 is a monomial; they cannot be canceled.

Also, 2x + 3 and x + 3 are not the *same* binomials; they cannot be canceled. The fraction

$\dfrac{2x+3}{2(x+3)}$ cannot be reduced.

(c) In the fraction $\dfrac{x}{x+3}$, x is a monomial and x + 3 is a binomial; they cannot be canceled.

The fraction $\dfrac{x}{x+3}$ cannot be reduced.

73

(4.1)

(d) In the fraction $\dfrac{x+2}{x^2+x+2}$, x + 2 is a binomial and $x^2 + x + 2$ is a trinomial; they cannot be factored

and they cannot be canceled. The fraction $\dfrac{x+2}{x^2+x+2}$ cannot be reduced.

EXAMPLE 5   Reduce the fraction $\dfrac{x^2y-4xy^2+4y^3}{x^2y-xy^2-6y^3}$.

$$\text{Answer:}\quad \frac{x^2y-4xy^2+4y^3}{x^2y-xy^2-6y^3}\underset{\text{Step (a)}}{=} \frac{y(x^2-4xy+4y^2)}{y(x^2+xy-6y^2)}\underset{\text{Step (b)}}{=} \frac{y(x-2y)(x-2y)}{y(x+3y)(x-2y)}\underset{\text{Step (c)}}{=} \frac{\cancel{y}(x-2y)(x\cancel{-2y})}{\cancel{y}(x+3y)(x\cancel{-2y})}= \frac{x-2y}{x+3y}$$

(a) Factor the numerator and the denominator using the Greatest Common Factor Method.
(b) Factor the numerator and the denominator again using the Trinomial Method.
(c) Then reduce the fraction and write the result as a separate step.

EXAMPLE 6   Reduce the fraction: $\dfrac{6-x}{x^2-36}$

✳ **Helpful Hint** – The numerator has subtraction with the terms in *reverse order*. The constant 6 is first and the variable x is the second term. (a) Rewrite the numerator in standard form (with the x term first). (b) Factor out −1 in the numerator. Factor the denominator using the Difference of Squares Method. (c) Then reduce the fraction and write the result.

$$\text{Answer:}\quad \frac{6-x}{x^2-36}\underset{\text{Step (a)}}{=} \frac{-x+6}{x^2-36}\underset{\text{Step (b)}}{=} \frac{-1(x-6)}{(x+6)(x-6)}\underset{\text{Step (c)}}{=} \frac{-1(x\cancel{-6})}{(x+6)(x\cancel{-6})}= -\frac{1}{x+6}$$

MULTIPLICATION AND DIVISION OF FRACTIONS

Multiplying Fractions

Factor each polynomial completely and cancel the common factors between the numerators and the denominators. Write the remaining factors as a separate step.

EXAMPLE 7   Simplify: $\dfrac{9x^2-9x}{3x^2-8x+5}\bullet\dfrac{9x^2-25}{18x^4+30x^3}$

$$\text{Answer:}\quad \frac{9x^2-9x}{3x^2-8x+5}\bullet\frac{9x^2-25}{18x^4+30x^3}= \underset{\text{Step (a)}}{\frac{9x(x-1)}{(3x-5)(x-1)}}\bullet\underset{\text{Step (b)}}{\frac{(3x-5)(3x+5)}{6x^3(3x+5)}}$$

$$= \underbrace{\frac{\overset{3}{\cancel{9x}}\cancel{(x-1)}}{\cancel{(3x-5)}\cancel{(x-1)}}\bullet\frac{\cancel{(3x-5)}\cancel{(3x+5)}}{\underset{2x^2}{\cancel{6x^3}}\cancel{(3x+5)}}}_{\text{Step (c)}}= \frac{3}{2x^2}$$

(a) In the first fraction, factor the numerator using the Greatest Common Factor Method.
   Factor the denominator using the Trinomial Method.
(b) In the second fraction, factor the numerator using the Difference of Squares Method.
   Factor the denominator using the Greatest Common Factor Method.
(c) Then cancel the common factors between numerators and denominators and write the result.

Division of Fractions

Change the division symbol to multiplication and invert the second fraction. Then factor each polynomial completely and cancel the common factors between numerators and denominators. Write the remaining factors as a separate step.

EXAMPLE 8   Simplify: $\dfrac{x^2-9y^2}{x^2+2xy} \div \dfrac{x^2-5xy+6y^2}{x^2-4y^2}$

Step (a)

Answer: $\dfrac{x^2-9y^2}{x^2+2xy} \div \dfrac{x^2-5xy+6y^2}{x^2-4y^2} = \dfrac{x^2-9y^2}{x^2+2xy} \cdot \dfrac{x^2-4y^2}{x^2-5xy+6y^2}$

Step (b)   Step (c)   Step (d)

$= \dfrac{(x-3y)(x+3y)}{x(x+2y)} \cdot \dfrac{(x-2y)(x+2y)}{(x-3y)(x-2y)} = \dfrac{\cancel{(x-3y)}\,(x+3y)}{x\,\cancel{(x+2y)}} \cdot \dfrac{\cancel{(x-2y)}\,\cancel{(x+2y)}}{\cancel{(x-3y)}\,\cancel{(x-2y)}} = \dfrac{x+3y}{x}$

(a) Change the operation symbol to multiplication and invert the second fraction.
(b) In the first fraction, factor the numerator using the Difference of Squares Method.
     Factor the denominator using the Greatest Common Factor Method.
(c) In the inverted second fraction, factor the numerator using the Difference of Squares Method.
     Factor the denominator using the Trinomial Method.
(d) Then cancel the common factors between numerators and denominators and write the result.

4.2   ADDITION AND SUBTRACTION OF FRACTIONS

SAME DENOMINATORS

To add or subtract fractions, each fraction must have the same denominator, which is called the **common denominator.**

Adding and Subtracting Fractions -- Same Denominators
(a) Bring down all of the numerators to one fraction bar. Bring down the common denominator.
(b) Add or subtract the similar terms of the numerator.
(c) If possible, factor polynomial in the numerator and in the denominator.
(d) Reduce the fraction *at the end of the problem.*

EXAMPLE 1   Simplify: $\dfrac{x+8}{x^2+3x} + \dfrac{2x+1}{x^2+3x}$

Step (a)   Step (b)   Step (c)   Step(d)

Answer: $\dfrac{x+8}{x^2+3x} + \dfrac{2x+1}{x^2+3x} = \dfrac{x+8+2x+1}{x^2+3x} = \dfrac{3x+9}{x^2+3x} = \dfrac{3(x+3)}{x(x+3)} = \dfrac{3\cancel{(x+3)}}{x\cancel{(x+3)}} = \dfrac{3}{x}$

EXAMPLE 2   Simplify: $\dfrac{y^2}{y^2+4y+3} - \dfrac{1}{y^2+4y+3}$

Steps (a & b)   Step (c)   Step (d)

Answer: $\dfrac{y^2}{y^2+4y+3} - \dfrac{1}{y^2+4y+3} = \dfrac{y^2-1}{y^2+4y+3} = \dfrac{(y-1)(y+1)}{(y+3)(y+1)} = \dfrac{(y-1)\cancel{(y+1)}}{(y+3)\cancel{(y+1)}} = \dfrac{y-1}{y+3}$

(4.2)

DIFFERENT DENOMINATORS

To add or subtract fractions which have different denominators, first set up the lowest common denominator for each fraction and set up new numerators.

The **lowest common denominator** (LCD) is the product of each different denominator factor.

EXAMPLE 3   State the LCD for $\dfrac{1}{5}$ and $\dfrac{x}{x+5}$.

Answer: The LCD is the product of the denominators. LCD $= 5(x+5)$.

Writing Fractions in Higher Terms

If you change the denominator of a fraction to the LCD, you must also *change the numerator*. Multiply the original fraction times a unit fraction:

$$\left(\frac{\text{missing factor}}{\text{missing factor}}\right)\left(\frac{\text{numerator}}{\text{denominator}}\right) = \frac{\text{new numerator}}{\text{LCD}}$$

EXAMPLE 4   For each fraction below, the denominator has been changed. Find the new numerator:

(a) $\dfrac{1}{5} = \dfrac{?}{5(x+5)}$
   (b) $\dfrac{x}{x+5} = \dfrac{?}{5(x+5)}$
   (c) $\dfrac{x+7}{x-3} = \dfrac{?}{(x-3)(x+3)}$

Answers:

(a) For $\dfrac{1}{5} = \dfrac{?}{5(x+5)}$, compare the first denominator 5 to the new denominator $5(x+5)$.

In the first denominator, the missing factor is $x+5$. Multiply $\dfrac{1}{5}$ times the unit fraction $\left(\dfrac{x+5}{x+5}\right)$ and

simplify the numerator: $\left(\dfrac{1}{5}\right)\left(\dfrac{x+5}{x+5}\right) = \dfrac{x+5}{5(x+5)}$

The new numerator is $x + 5$.

(b) For $\dfrac{x}{x+5} = \dfrac{?}{5(x+5)}$, compare the first denominator $x + 5$ to the new denominator $5(x+5)$.

In the first denominator, the missing factor is 5. Multiply the unit fraction $\left(\dfrac{5}{5}\right)$ times $\dfrac{x}{x+5}$ and

simplify the numerator: $\left(\dfrac{5}{5}\right)\left(\dfrac{x}{x+5}\right) = \dfrac{5x}{5(x+5)}$

The new numerator is $5x$.

(c) For $\dfrac{x+7}{x-3} = \dfrac{?}{(x-3)(x+3)}$, compare the first denominator $x - 3$ to the denominator $(x-3)(x+3)$.

In the first denominator, the missing factor is $x + 3$. Multiply $\dfrac{x+7}{x-3}$ times the unit fraction $\left(\dfrac{x+3}{x+3}\right)$,

simplify the numerator using "F.O.I.L.":

$$\left(\frac{x+7}{x-3}\right)\left(\frac{x+3}{x+3}\right) = \frac{x^2+3x+7x+21}{(x+3)(x-3)} = \frac{x^2+10x+21}{(x+3)(x-3)}. \quad \text{The new numerator is } x^2+10x+21.$$

Adding and Subtracting Fractions -- Different Denominators

(a) Set up the *LCD* for each fraction.
(b) Set up the *new numerators.*
(c) *Bring down* all of the numerators to one fraction bar. Bring down the LCD.
(d) Add or subtract the *similar terms* of the numerator. Bring down the LCD.
(e) *Factor and reduce* the fraction at the end of the problem, if possible.

EXAMPLE 5 Simplify: $\frac{1}{5} + \frac{x}{x+5}$

Answer:

(a) Set up the *LCD*: The LCD is the product of the different denominators. LCD = 5(x + 5)

(b) Set up the *new numerators.* $\frac{1}{5} + \frac{x}{x+5} = \frac{1}{5}\left(\frac{x+5}{x+5}\right) + \left(\frac{5}{5}\right)\frac{x}{x+5} = \frac{x+5}{5(x+5)} + \frac{5x}{5(x+5)}$

(c) Bring down all of the numerators to one fraction bar. Bring down the LCD. $= \frac{x + 5 + 5x}{5(x+5)}$

(d) Add the *similar terms* of the numerator.

Bring down the LCD. $= \frac{6x+5}{5(x+5)}$

(e) The numerator cannot be factored and the fraction cannot be reduced.

Helpful Hint – If a monomial denominator has an exponent, the LCD will include the exponent. The lower exponent term is a factor of a higher exponent term. In other words, x is a factor of x^2. Also x and x^2 are factors of x^3; x, x^2, and x^3 are factors of x^4, etc.

EXAMPLE 6 Simplify: $\frac{x+2}{3x} - \frac{x+1}{x^2}$

Answer: In this example, the LCD is $3x^2$. Use the steps for adding and subtracting fractions.

(a) Set up the *LCD* for each fraction. $\frac{x+2}{3x} - \frac{x+1}{x^2} = \frac{?}{3x^2} - \frac{??}{3x^2}$

(b) Set up the *new numerators.* $= \frac{x+2}{3x}\left(\frac{x}{x}\right) - \left(\frac{3}{3}\right)\frac{x+1}{x^2} = \frac{x^2+2x}{3x^2} - \frac{3x+3}{3x^2}$

(c) Bring down all of the numerators to one fraction bar.

Bring down the LCD. $= \frac{x^2+2x-3x-3}{3x^2}$

(d) Subtract the *similar terms* of the numerator.

Bring down the LCD. $= \frac{x^2-x-3}{3x^2}$

(e) The numerator cannot be factored and the fraction cannot be reduced.

(4.2)

EXAMPLE 7   Simplify: $\dfrac{x+7}{x-3} - \dfrac{x+6}{x+3}$

Answer: The LCD is $(x-3)(x+3)$. Use the steps for adding and subtracting fractions.

(a) Set up the *LCD* for each fraction.

$$\frac{x+7}{x-3} - \frac{x+6}{x+3} = \frac{?}{(x-3)(x+3)} - \frac{??}{(x-3)(x+3)}$$

(b) Set up the *new numerators*.

$$= \frac{x+7}{x-3}\left(\frac{x+3}{x+3}\right) - \left(\frac{x-3}{x-3}\right)\frac{x+6}{x+3} = \frac{x^2+3x+7x+21}{(x-3)(x+3)} - \frac{x^2+6x-3x-18}{(x-3)(x+3)}$$

(c) Bring down all of the numerators to one fraction bar.

Bring down the LCD.

$$= \frac{x^2+10x+21}{(x-3)(x+3)} - \frac{x^2+3x-18}{(x-3)(x+3)} = \frac{x^2+10x+21-x^2-3x+18}{(x-3)(x+3)}$$

(d) Add or subtract the *similar terms* of the numerator.

Bring down the LCD.

$$= \frac{7x+39}{(x-3)(x+3)}$$

(e) The fraction cannot be reduced.

EXAMPLE 8   Simplify: $\dfrac{x}{x^2-9} - \dfrac{2}{x^2-2x-3}$

Answer: To determine the LCD, factor both denominators.

$$\frac{x}{x^2-9} - \frac{2}{x^2-2x-3} = \frac{x}{(x-3)(x+3)} - \frac{2}{(x-3)(x+1)}$$

The LCD is the product of each different denominator factor. (Do *not* repeat a factor.)
For this problem, the LCD has 3 factors: LCD = $(x-3)(x+3)(x+1)$.

(a) Set up the *LCD* for each fraction.

$$\frac{x}{(x-3)(x+3)} - \frac{2}{(x-3)(x+1)}$$

$$= \frac{?}{(x-3)(x+3)(x+1)} - \frac{??}{(x-3)(x+3)(x+1)}$$

(b) Set up the *new numerators*.

$$= \frac{x}{(x-3)(x+3)}\left(\frac{x+1}{x+1}\right) - \frac{2}{(x-3)(x+1)}\left(\frac{x+3}{x+3}\right)$$

$$= \frac{x^2+x}{(x-3)(x+3)(x+1)} - \frac{2x+6}{(x-3)(x+3)(x+1)}$$

(c) Bring down all of the numerators to one fraction bar.

Bring down the LCD.

$$= \frac{x^2+x-2x-6}{(x-3)(x+3)(x+1)}$$

(d) Add or subtract the *similar terms* of the numerator.

Bring down the LCD.

$$= \frac{x^2-x-6}{(x-3)(x+3)(x+1)}$$

(e) Factor the numerator and reduce the fraction.

$$= \frac{\cancel{(x-3)}(x+2)}{\cancel{(x-3)}(x+3)(x+1)} = \frac{x+2}{(x+3)(x+1)}$$

4.3   COMPLEX FRACTIONS

A **complex fraction** has a fraction within a fraction. We will rewrite a complex fraction as one fraction by multiplying the numerator and denominator times the lowest common denominator (LCD):

$$\frac{Complex}{Fraction}\left(\frac{LCD}{LCD}\right) = \frac{One}{Fraction}$$

NUMERICAL COMPLEX FRACTIONS

EXAMPLE 1   Simplify the complex fraction: $\dfrac{\dfrac{2}{3}+\dfrac{1}{8}}{\dfrac{1}{4}-\dfrac{5}{6}}$.

Answer:

The lowest common denominator is 24. (a) Multiply each term of the complex fraction times 24. (b) For each term of the complex fraction, cancel out the denominator and multiply the result times the numerator. The result will be one fraction. (c) Simplify the numerator and simplify the denominator.

$$\frac{\frac{2}{3}+\frac{1}{8}}{\frac{1}{4}-\frac{5}{6}} = \frac{\frac{2}{3}+\frac{1}{8}}{\frac{1}{4}-\frac{5}{6}}\cdot\left[\frac{\frac{24}{1}}{\frac{24}{1}}\right] = \frac{\frac{2}{3}\left(\frac{24}{1}\right)+\frac{1}{8}\left(\frac{24}{1}\right)}{\frac{1}{4}\left(\frac{24}{1}\right)-\frac{5}{6}\left(\frac{24}{1}\right)} = \frac{2(8)+1(3)}{1(6)-5(4)} = \frac{16+3}{6-20} = -\frac{19}{14}$$

Step (a)   Step (b)   Step (c)

COMPLEX FRACTIONS WITH VARIABLE TERMS

EXAMPLE 2   **Monomial Denominators:** Simplify the complex fraction $\dfrac{1-\dfrac{4}{x^2}}{1+\dfrac{3}{x}+\dfrac{2}{x^2}}$.

Answer:

Remember that x is a factor of x^2. Therefore the lowest common denominator is x^2. (a) Multiply each term of the complex fraction times x^2. The result will be one fraction. (b) Factor the numerator using the Difference of Squares Method. Factor the denominator using the Trinomial Method. (c) Reduce the fraction.

Step (a)

$$\frac{1-\frac{4}{x^2}}{1+\frac{3}{x}+\frac{2}{x^2}} = \frac{1-\frac{4}{x^2}}{1+\frac{3}{x}+\frac{2}{x^2}}\cdot\left[\frac{\frac{x^2}{1}}{\frac{x^2}{1}}\right] = \frac{1\left(\frac{x^2}{1}\right)-\frac{4}{x^2}\left(\frac{x^2}{1}\right)}{1\left(\frac{x^2}{1}\right)+\frac{3}{x}\left(\frac{x^2}{1}\right)+\frac{2}{x^2}\left(\frac{x^2}{1}\right)}$$

$$= \frac{x^2-4}{x^2+3x+2} = \frac{(x-2)(x+2)}{(x+1)(x+2)} = \frac{x-2}{x+1}$$

Step (b)   Step (c)

79

(4.3)

EXAMPLE 3   **Polynomial Denominators:** Simplify the complex fraction $\dfrac{\dfrac{x}{x+6} - \dfrac{x}{x-6}}{\dfrac{x}{x+6} + \dfrac{x}{x-6}}$.

Answer:

(a) The lowest common denominator is $(x + 6)(x - 6)$. Multiply each term of the complex fraction times $(x + 6)(x - 6)$. (b) For each term of the complex fraction, cancel out the denominator and multiply the result times the numerator. The result will be one fraction. (c) Simplify the numerator. Simplify the denominator. (d) Reduce the fraction.

$$\frac{\dfrac{x}{x+6} - \dfrac{x}{x-6}}{\dfrac{x}{x+6} + \dfrac{x}{x-6}} = \frac{\dfrac{x}{x+6} - \dfrac{x}{x-6}}{\dfrac{x}{x+6} + \dfrac{x}{x-6}} \cdot \left[\frac{\dfrac{(x+6)(x-6)}{1}}{\dfrac{(x+6)(x-6)}{1}} \right]$$

Step (a)

$$= \frac{\dfrac{x}{x+6} \cdot \left[\dfrac{(x+6)(x-6)}{1}\right] - \dfrac{x}{x-6} \cdot \left[\dfrac{(x+6)(x-6)}{1}\right]}{\dfrac{x}{x+6} \cdot \left[\dfrac{(x+6)(x-6)}{1}\right] + \dfrac{x}{x-6} \cdot \left[\dfrac{(x+6)(x-6)}{1}\right]}$$

Step (b)   Step (c)   Step (d)

$$= \frac{x(x-6) - x(x+6)}{x(x-6) + x(x+6)} = \frac{x^2 - 6x - x^2 - 6x}{x^2 - 6x + x^2 + 6x} = \frac{-12x}{2x^2} = -\frac{6}{x}$$

4.4   RATIONAL EQUATIONS (EQUATIONS WITH FRACTIONS)

A **rational equation** has one or more fractional terms. To solve a rational equation, *eliminate the fractions* by multiplying each term times the lowest common denominator (LCD). Rewrite the equation and solve it as illustrated in the following examples.

RATIONAL EQUATION WITH NUMERICAL DENOMINATORS

EXAMPLE 1   Solve the equation: $\dfrac{x}{6} + \dfrac{5x-1}{3} = \dfrac{7}{4}$

Answer:

The LCD is 12. Multiply each term times 12.

For each term, cancel out the denominator.
Multiply the result times the numerator.

Combine the similar terms.

Solve the equation for x.

$$(12)\frac{x}{6} + (12)\frac{5x-1}{3} = (12)\frac{7}{4}$$

$$2x + 20x - 4 = 21$$

$$22x - 4 = 21$$

$$22x = 25$$

$$x = \frac{25}{22}$$

80

RATIONAL EQUATION WITH A VARIABLE IN THE DENOMINATOR

In Section 4.1, we observed that the denominator of a fraction cannot equal zero. If a variable is in the denominator, state the restricted value(s) and solve the equation using the following procedure.

Steps:

- State the *restricted value(s)* for each denominator which has a variable.
- *Multiply the LCD* times each term of the equation.
- For each term, *cancel out the denominator* and multiply the result times the numerator.
- Rewrite the equation. *Solve* this equation.
- *Compare the results* to the restricted value(s). If a restricted value is one of the results, delete it from the solution set.

Helpful Hint *–A restricted value cannot be given as a solution to a rational (fractional) equation.*

EXAMPLE 2 Solve the equation: $\frac{3}{x-3} + \frac{2}{x} = \frac{2}{x-3}$

State the restricted values. In this equation, $x - 3 \neq 0$ and $x \neq 0$. Restricted values: $x \neq 3, x \neq 0$

Multiply the LCD, $x(x - 3)$, times each term of the equation.

$$x(x-3)\frac{3}{x-3} + x(x-3)\frac{2}{x} = x(x-3)\frac{2}{x-3}$$

For each term, cancel out the denominator. Multiply the result times the numerator.

$$(x)\cancel{(x-3)}\frac{3}{\cancel{x-3}} + \cancel{(x)}(x-3)\frac{2}{\cancel{x}} = (x)\cancel{(x-3)}\frac{2}{\cancel{x-3}}$$

Rewrite the equation. $(x)(3) + (x-3)(2) = (x)(2)$

Combine the similar terms. Solve the equation. $3x + 2x - 6 = 2x$

Subtract 5x on both sides of the equation. $5x - 6 = 2x$

Divide by -3 on both sides of the equation. $-6 = -3x$

Compare the result to the restricted values. $2 = x$

(The result is not one of the restricted values.) Answer: $x = 2$

EXAMPLE 3 Solve the equation: $x + \frac{7}{x} = \frac{16}{3}$

State the restricted value. $x \neq 0$

Multiply the LCD 3x times each term of the equation. $(3x)\,x + (3x)\frac{7}{x} = (3x)\frac{16}{3}$

For each term, cancel out the denominator. Multiply the result times the numerator.

$$(3x)\,x + (3\cancel{x})\frac{7}{\cancel{x}} = (\cancel{3}x)\frac{16}{\cancel{3}}$$

(4.4)

Rewrite the equation. $(3x)(x) + (3)(7) = (x)(16)$

This equation has an exponent. Therefore, the equation $3x^2 + 21 = 16x$
must equal zero. Subtract 16x on both sides of the equation.

$$3x^2 - 16x + 21 = 0$$

Factor the polynomial using the Trinomial Method. $(3x - 7)(x - 3) = 0$

Solve each equation. $3x - 7 = 0$ $\quad\bigg|\quad$ $x - 3 = 0$

Compare the results to the restricted value. $3x = 7$

The equation has two solutions.   Answers: $x = \dfrac{7}{3}$ $\quad\bigg|\quad$ $x = 3$

EXAMPLE 4   Solve the equation: $\dfrac{2}{x^2 - 2x} - \dfrac{1}{x - 2} = \dfrac{2}{x}$

Factor the first denominator using the Greatest Common Factor Method. $\dfrac{2}{x(x - 2)} - \dfrac{1}{x - 2} = \dfrac{2}{x}$

State the restricted values. $x \neq 0,\ x \neq 2$

Multiply the LCD x(x – 2) times each term of the equation.

$$(x)(x - 2)\dfrac{2}{x(x - 2)} - (x)(x - 2)\dfrac{1}{x - 2} = (x)(x - 2)\dfrac{2}{x}$$

For each term, cancel out the denominator. Multiply the result times the numerator.

$$(x)(x - 2)\dfrac{2}{x(x - 2)} - (x)(x - 2)\dfrac{1}{x - 2} = (x)(x - 2)\dfrac{2}{x}$$

Rewrite the equation. Solve this equation. $2 - (x)(1) = (x - 2)(2)$

Simplify both sides of the equation. $2 - x = 2x - 4$

Subtract 2x on both sides of the equation. $2 - 3x = -4$

Subtract 2 on both sides of the equation. $-3x = -6$

Divide by – 3 on both sides of the equation. $x = 2$

Compare the result to the restricted value.   2 is a restricted value.
                                                              Delete this result.

Answer: No solution

EXAMPLE 5 Solve the equation $\frac{x}{x^2-25} - \frac{4}{x^2-2x-15} = 0$

Factor the first denominator using the Difference of Squares Method.

Factor the second denominator using the Trinomial Method. $\frac{x}{(x+5)(x-5)} - \frac{4}{(x-5)(x+3)} = 0$

State the restricted values. $x \neq -5,\ x \neq 5,\ x \neq -3$

Multiply the LCD (x + 5)(x – 5)(x + 3) times each term of the equation.

$$(x+5)(x-5)(x+3)\frac{x}{(x+5)(x-5)} - (x+5)(x-5)(x+3)\frac{4}{(x-5)(x+3)} = 0(x+5)(x-5)(x+3)$$

For each term, cancel out the denominator. Multiply the result times the numerator.

$$(x+5)(x-5)(x+3)\frac{x}{(x+5)(x-5)} - (x+5)(x-5)(x+3)\frac{4}{(x-5)(x+3)} = 0(x+5)(x-5)(x+3)$$

Rewrite the equation. $(x+3)(x) - (4)(x+5) = 0$

Combine the similar terms. $x^2 + 3x - 4x - 20 = 0$

$x^2 - x - 20 = 0$

Factor the polynomial using the Trinomial Method. $(x-5)(x+4) = 0$

Solve each equation. $x - 5 = 0$ | $x + 4 = 0$

Compare the result to the restricted value. $x = 5$ | $x = -4$

5 is a restricted value. Delete this result. | This answer is acceptable.

Answer: $x = -4$

Helpful Hint – Summary (A Comparison of Sections 4.4 and 4.2)

Remember that solving a fractional equation is _not_ the same as adding or subtracting fractions.

For An Equation (4.4): ***Equal Sign*** ⟹ Solve the equation the **"e a s y w a y."**

EXAMPLE 6 Solve: $\frac{5}{x-5} = \frac{2}{x} + \frac{2}{x-5}$

State the restricted values. $x \neq 5,\ x \neq 0$

Write the LCD (x – 5)(x) at the ***top*** of each term. $(x-5)(x)\frac{5}{x-5} = (x-5)(x)\frac{2}{x} + (x-5)(x)\frac{2}{x-5}$

Cancel out the denominators. $(x-5)(x)\frac{5}{x-5} = (x-5)(x)\frac{2}{x} + (x-5)(x)\frac{2}{x-5}$

Rewrite the equation and solve it. $5x = 2x - 10 + 2x$

$5x = 4x - 10$

The result is not one of the restricted values. Answer: $x = -10$

(4.4)

For Addition or Subtraction (from Section 4.2):

No equal sign ⟹ Work the problem the **"longer way."**
Write the LCD at the **bottom** of each term.
Show the LCD for each step.

EXAMPLE 7

Simplify: $\dfrac{5}{x-5} + \dfrac{2}{x} + \dfrac{2}{x-5}$

Answer: The LCD is $(x-5)(x)$.

(a) Set up the LCD for each fraction.

$$= \frac{?}{(x-5)(x)} + \frac{??}{(x-5)(x)} + \frac{???}{(x-5)(x)}$$

(b) Set up the new numerators.

$$= \frac{5}{x-5}\left(\frac{x}{x}\right) + \left(\frac{x-5}{x-5}\right)\left(\frac{2}{x}\right) + \frac{2}{x-5}\left(\frac{x}{x}\right)$$

$$= \frac{5x}{(x-5)(x)} + \frac{2x-10}{(x-5)(x)} + \frac{2x}{(x-5)(x)}$$

(c) Bring down all of the numerators to one fraction bar.

Bring down the LCD.

$$= \frac{5x+2x-10+2x}{x(x-5)}$$

(d) Add the similar terms of the numerator.

Bring down the LCD.

Answer: $= \dfrac{9x-10}{x(x-5)}$

(e) The numerator cannot be factored and the fraction cannot be reduced.

*Observe that the answers for Example 6 and Example 7 are **not** the same.*

4.5   APPLICATIONS FOR RATIONAL EQUATIONS

NUMBER PROBLEMS -- WORKING WITH RECIPROCALS

To write the reciprocal of a given number, invert the number.

Number	Reciprocal	
3	$\dfrac{1}{3}$	
$-\dfrac{1}{2}$	$-\dfrac{2}{1} = -2$	
$\dfrac{3}{4}$	$\dfrac{4}{3}$	
x	$\dfrac{1}{x}$	$(x \neq 0)$
$x+2$	$\dfrac{1}{x+2}$	$(x \neq -2)$

(4.5)

EXAMPLE 1   Write the equation and solve:

The sum of the reciprocals of two consecutive even integers is $\dfrac{5}{12}$. Find the numbers.

�֎ ___Helpful Hint___ – (1) To review the set-up of consecutive even integers, you may refer to Section 1.3, Example 4. (2) Set up an integer column and a reciprocal column. Use the reciprocal column to write the equation. Solve the equation and find the answers for the integer column.

Table	(Integer)	(Reciprocal)
1st integer	x	$\dfrac{1}{x}$
2nd integer	x + 2	$\dfrac{1}{x+2}$

Add the reciprocals to write the equation.

$$\frac{1}{x} + \frac{1}{x+2} = \frac{5}{12}$$

In this equation, $x \neq 0$ and $x + 2 \neq 0$. Restricted values: $x \neq 0, x \neq -2$

Multiply the LCD, $(12)(x)(x + 2)$, times each term of the equation.

$$(12)(x)(x+2)\frac{1}{x} + (12)(x)(x+2)\frac{1}{x+2} = \frac{5}{12}(12)(x)(x+2)$$

For each term, cancel out the denominator. Multiply the result times the numerator.

$$(12)\,\cancel{(x)}\,(x+2)\frac{1}{\cancel{x}} + (12)\,(x)\,\cancel{(x+2)}\frac{1}{\cancel{x+2}} = \frac{5}{\cancel{12}}\cancel{(12)}(x)\,(x+2)$$

$$(12)(x + 2)(1) + (12x)(1) = (5x)(x + 2)$$

$$12x + 24 + 12x = 5x^2 + 10x$$

Combine the similar terms on the left side.

$$24x + 24 = 5x^2 + 10x$$

This equation has an exponent; the equation must equal 0.
The leading term $5x^2$ is positive.
Cancel the left side of the equation.

$$\underline{-24x - 24 \qquad\qquad -24x - 24}$$

$$0 = 5x^2 - 14x - 24$$

Factor completely (using Trinomial Factoring).

$$0 = (x - 4)(5x + 6)$$

Set each variable factor equal to 0.
Solve for each *x*.

$x - 4 = 0$	$5x + 6 = 0$
$x = 4$	$5x = -6$

Each answer must be an integer.

Recall that an integer is a positive or negative whole number, or zero.

	$x = -\dfrac{\cancel{6}}{5}$
$x + 2 =$	A fraction is not an integer.
$4 + 2 = 6$	Delete this result.

Answers: The numbers are 4 and 6.

To check the answers, add the reciprocals:

$$\frac{1}{4} + \frac{1}{6} = \frac{1}{4}\left(\frac{3}{3}\right) + \frac{1}{6}\left(\frac{2}{2}\right) = \frac{3}{12} + \frac{2}{12} = \frac{5}{12}$$

(4.5)

WORK PROBLEMS

Two machines, or two people, or two pipes, etc., are working together the finish the same job.

Work Formula:

$$\frac{\text{rate together}}{\text{1st rate}} \ + \ \frac{\text{rate together}}{\text{2nd rate}} \ = \ 1 \qquad \text{(One completed job)}$$

In this formula, the first fraction represents the fractional part of the job done by the first person or machine. The second fraction represents the fractional part of the job done by the second person or machine. The "1" represents one completed job.

EXAMPLE 2   Ron can wallpaper a room in 10 hours. It takes his wife Rita only 8 hours to do the same job. If they work together, how long will it take them to wallpaper the room?

Workers	Rates
Ron	10 hrs
Rita	8 hrs
Together	x

Use the Work Formula to write the equation. $\dfrac{x}{10} + \dfrac{x}{8} = 1$

The denominators are numbers. There are no restricted values.

Multiply the LCD (40) times each term of the equation. $(40)\dfrac{x}{10} \ + \ (40)\dfrac{x}{8} \ = \ 1\,(40)$

Cancel each denominator. $\overset{(4)}{\cancel{(40)}}\dfrac{x}{\cancel{10}} \ + \ \overset{(5)}{\cancel{(40)}}\dfrac{x}{\cancel{8}} \ = \ 1\,(40)$

Multiply the result times the numerator. $4x + 5x = 40$

Combine the similar terms. $9x = 40$

Solve for x. $x = \dfrac{40}{9} = 4\dfrac{4}{9} \ hrs$

Answer: Working together, it will take Ron and Rita $4\dfrac{4}{9}$ hours to paint the room.

EXAMPLE 3   It takes Stan 9 hours longer to wash the windows of a building than it takes Mark. Working together, Mark and Stan can wash the windows of the building in 6 hours. How long does it take for Stan to do the job if he works alone?

Workers	Rates
Stan	x + 9
Mark	x
Together	6 hrs

(4.5)

Use the Work Formula to write the equation.

$$\frac{6}{x+9} + \frac{6}{x} = 1$$

In this equation, $x + 9 \neq 0$ and $x \neq 0$. Restricted values: $x \neq -9, x \neq 0$

Multiply the LCD $(x + 9)(x)$ times each term of the equation.

$$(x+9)(x)\frac{6}{x+9} + (x+9)(x)\frac{6}{x} = 1(x+9)(x)$$

Cancel each denominator.

$$(x+9)(x)\frac{6}{x+9} + (x+9)(x)\frac{6}{x} = 1(x+9)(x)$$

Multiply the result times the numerator.

$$6x + 6x + 54 = x^2 + 9x$$

Combine the similar terms.

$$12x + 54 = x^2 + 9x$$

The equation has an exponent; the equation must equal zero.
The leading term x^2 is positive.

$$-12x - 54 \qquad -12x - 54$$

Cancel the left side of the equation.

$$0 = x^2 - 3x - 54$$

Factor completely (using Trinomial Factoring).

$$0 = (x - 9)(x + 6)$$

Solve for x.

$x - 9 = 0$ $\qquad\qquad$ $x + 6 = 0$

Answers: $\qquad\qquad\qquad$ $x = 9$ hrs   (Mark) $\qquad$ $x = -6$

$\qquad\qquad\qquad\qquad$ $x + 9 = 18$ hrs (Stan) $\qquad$ Time is positive.
Delete this answer.

✠ **_Helpful Hint_** - The Work Formula, $\dfrac{\text{rate together}}{\text{1st rate}} + \dfrac{\text{rate together}}{\text{2nd rate}} = 1$, can be easily modified based on
the information given in the word problem. The "1" represents one complete job. If part of the job
was completed, change the 1 to the fraction or decimal part of the job that was finished.

EXAMPLE 4 $\qquad$ A hot water faucet can fill a tub in 18 minutes. The cold water faucet can fill the tub
in 12 minutes. How long will it take to fill three-fourths of the tub if both faucets are turned on?

Faucets $\qquad\qquad$ Rates
Hot $\qquad\qquad\qquad$ 18 min
Cold $\qquad\qquad\quad$ 8 hrs
Together $\qquad\qquad$ x

Using the Work Formula, change 1 to $\frac{3}{4}$. $\qquad\qquad$ ("Fill three-fourths of the tub")

Write the equation.

$$\frac{x}{18} + \frac{x}{12} = \frac{3}{4}$$

The denominators are numbers. There are no restricted values.

Multiply the LCD (36) times each term of the equation. $\qquad (36)\frac{x}{18} + (36)\frac{x}{12} = (36)\frac{3}{4}$

Cancel each denominator.

$$\overset{(2)}{(36)}\frac{x}{18} + \overset{(3)}{(36)}\frac{x}{12} = \overset{(9)}{(36)}\frac{3}{4}$$

87

(4.5)

Multiply the result times the numerator.	$2x + 3x = 27$
Combine the similar terms.	$5x = 27$
Solve for x.	$x = \dfrac{27}{5} = 5.4$ min

Answer: It will take 5.4 minutes to fill the tub.

✴ *Helpful Hint* – Inlet and Outlet Pipes:

Using the Work Formula, $\dfrac{\text{rate together}}{\text{1st rate}} + \dfrac{\text{rate together}}{\text{2nd rate}} = 1$, let the first fraction represent the inlet pipe and let the second fraction represent the outlet pipe. Remember that an inlet pipe lets water *in*; the first fraction will be *positive*. An outlet pipe lets water <u>*out*</u>; change the sign of the second fraction to *negative*. The following example shows how to solve this type of word problem.

EXAMPLE 5   An inlet pipe can fill a water tank in 12 hours. An outlet pipe can empty the tank in 20 hours. How long will it take to fill the tank if the inlet and outlet pipes are left open?

Pipes	Rates
Inlet	12 hrs
Outlet	20 hrs
Both open together	x

Using the Work Formula: The first fraction, representing the inlet pipe, will be positive. The second fraction, representing the outlet pipe, will be *negative*.

Write the equation.

$$\underset{\text{Inlet}}{\dfrac{x}{12}} - \underset{\text{Outlet}}{\dfrac{x}{20}} = \underset{\text{One full tank}}{1}$$

The denominators are numbers. There are no restricted values.

Multiply the LCD (60) times each term of the equation.

$$(60)\dfrac{x}{12} - (60)\dfrac{x}{20} = 1(60)$$

Cancel each denominator.

$$\overset{(5)}{\cancel{(60)}\dfrac{x}{\cancel{12}}} - \overset{(3)}{\cancel{(60)}\dfrac{x}{\cancel{20}}} = 1(60)$$

Multiply the result times the numerator.	$5x - 3x = 60$
Combine the similar terms.	$2x = 60$
Solve for x.	$x = 30$ hrs

Answer: If the inlet and outlet pipes are left open, it will take 30 hours to fill the tank.

Chapter 5   **Radicals and Equations**

5.1   ROOTS AND RADICALS

A **radical** is a mathematical expression which contains a "root" symbol $\sqrt[N]{}$. For any radical, the number on the upper left is called the **index**. The expression under the radical symbol is the **radicand**.

$$\text{index} \longrightarrow \sqrt[N]{\underbrace{XY}_{\text{radicand}}}$$

If the index is not given, the radical is a square root: $\sqrt{X} = \sqrt[2]{X}$.

The expression, $\sqrt[4]{X}$ is called the "4th root of x". The index is 4 and the radicand is x.

EXAMPLE 1   Simplify each of the following radicals:

(a) $\sqrt{4}$ (b) $\sqrt{9}$ (c) $\sqrt[3]{27}$ (d) $\sqrt[3]{64}$ (e) $\sqrt[4]{10,000}$ (f) $\sqrt[5]{32}$

Answers:

(a) $\sqrt{4}$ means "What number squared equals 4?" Given $2^2 = 4$, the radical $\sqrt{4} = 2$.

(b) Given $3^2 = 9$, the radical $\sqrt{9} = 3$.

(c) $\sqrt[3]{27}$ means "What number cubed equals 27?" Given $3^3 = 27$, the radical $\sqrt[3]{27} = 3$.

(d) Given $4^3 = 64$, the radical $\sqrt[3]{64} = 4$.

(e) $\sqrt[4]{10,000}$ means "What number to the 4th power equals 10,000?" Given $10^4 = 10,000$ the radical $\sqrt[4]{10,000} = 10$.

(f) $\sqrt[5]{32}$ means "What number to the 5th power equals 32?" Given $2^5 = 32$, the radical $\sqrt[5]{32} = 2$.

THE EVEN INDEX $\sqrt{}$, $\sqrt[4]{}$, $\sqrt[6]{}$, $\ldots$

To get a real number answer, *the radicand must be positive or zero.*

EXAMPLE 2   Simplify each of the following radicals:

(a) $\sqrt{25}$ (b) $-\sqrt{25}$ (c) $\sqrt{-25}$ (d) $\sqrt{0}$ (e) $\sqrt{1}$

Answers:

(a) $\sqrt{25} = 5$ (b) $-\sqrt{25} = -(5) = -5$

(c) $\sqrt{-25}$ is not a real number. It is an imaginary number. (Refer to Section 5.7 for details.)

(d) $\sqrt{0}$ must be simplified; $\sqrt{0} = 0$.   (e) $\sqrt{1}$ must be simplified; $\sqrt{1} = 1$.

(5.1)

THE ODD INDEX $\sqrt[3]{}$, $\sqrt[5]{}$, $\sqrt[7]{}$, $\cdots$

The radicand may be positive, negative, or zero. The answer is always a real number. If the radicand is positive, the answer is positive. If the radicand is negative, the answer is negative.

EXAMPLE 3   Simplify each of the following radicals:

(a) $\sqrt[3]{8}$ (b) $\sqrt[3]{-8}$ (c) $\sqrt[5]{32}$ (d) $\sqrt[5]{-32}$ (e) $\sqrt[7]{1}$ (f) $\sqrt[7]{-1}$

Answers:

(a) $\sqrt[3]{8} = 2$ (b) $\sqrt[3]{-8} = -2$ (c) $\sqrt[5]{32} = 2$ (d) $\sqrt[5]{-32} = -2$

(e) $\sqrt[7]{1} = 1$ (f) $\sqrt[7]{-1} = -1$

REDUCING RADICALS INVOLVING VARIABLES   (Part 1)*

REDUCTION RULE: Write the root of the numerical coefficient. Divide the index into the exponent of each variable.

EXAMPLE 4   Simplify each of the following radicals: (a) $\sqrt{25x^2}$ (b) $\sqrt[3]{8a^3b^9}$
(c) $\sqrt[5]{-32y^{10}}$ (d) $\sqrt[4]{16m^{12}p^{20}}$ (e) $\sqrt[3]{125a^{12}b^{21}}$ (f) $\sqrt[7]{-x^{35}y^7z^{14}}$

Answers:

(a) For $\sqrt{25x^2}$, the index is 2. The square root of 25 is 5. Divide index 2 into the exponent.
The result is $\sqrt{25x^2} = 5x$.

(b) For $\sqrt[3]{8a^3b^9}$, the cube root of 8 is 2. Divide index 3 into each exponent. The result is
$\sqrt[3]{8a^3b^9} = 2ab^3$.

(c) For $\sqrt[5]{-32y^{10}}$, the fifth root of -32 is -2. Divide index 5 into the exponent. The result is
$\sqrt[5]{-32y^{10}} = -2y^2$.

(d) For $\sqrt[4]{16m^{12}p^{20}}$, the fourth root of 16 is 2. Divide index 4 into each exponent. The result is
$\sqrt[4]{16m^{12}p^{20}} = 2m^3p^5$.

(e) For $\sqrt[3]{125a^{12}b^{21}}$, the cube root of 125 is 5. Divide index 3 into each exponent. The result is
$\sqrt[3]{125a^{12}b^{21}} = 5a^4b^7$.

(f) For $\sqrt[7]{-x^{35}y^7z^{14}}$, the seventh root of -1 is -1. Divide index 7 into each exponent. The result is
$\sqrt[7]{-x^{35}y^7z^{14}} = -x^5yz^2$.

* ✖ *Helpful Hint* –For "**Reducing Radicals, Part (II)**", refer to Section 5.3 of this Study Guide.

5.2   RATIONAL EXPONENTS

CHANGING A RATIONAL EXPONENT TO RADICAL FORM

A fractional or decimal exponent is called a rational exponent. An expression which has a rational exponent represents a radical.

PROPERTY 1

> If N is a positive integer and $\sqrt[N]{A}$ is a real number
>
> then $A^{\frac{1}{N}} = \sqrt[N]{A}$.

Observe that the denominator becomes the index.

$A^{\frac{1}{2}}$ means square root of A. $A^{\frac{1}{3}}$ means cube root of A.

$A^{\frac{1}{4}}$ means 4th root of A. $A^{\frac{1}{5}}$ means 5th root of A.

EXAMPLE 1   Change to radical form and simplify:

(a) $64^{\frac{1}{2}}$ (b) $1000^{\frac{1}{3}}$ (c) $121^{\frac{1}{2}}$ (d) $(-125)^{\frac{1}{3}}$

Answers:

(a) $64^{\frac{1}{2}} = \sqrt{64} = 8$ (b) $1000^{\frac{1}{3}} = \sqrt[3]{1000} = 10$

(c) $16^{\frac{1}{4}} = \sqrt[4]{16} = 2$ (d) $(-32)^{\frac{1}{5}} = \sqrt[5]{-32} = -2$

MORE RATIONAL EXPONENTS   (Without using a calculator)

PROPERTY 2

> If M and N are positive integers and $A^{\frac{1}{N}}$ is a real number
>
> then $A^{\frac{M}{N}} = \left(A^{\frac{1}{N}} \right)^{M}$.

✖ *Helpful Hint* – Move the numerator outside of the parentheses.

EXAMPLE 2   Simplify without using a calculator:

(a) $9^{\frac{3}{2}}$ (b) $32^{\frac{2}{5}}$ (c) $16^{\frac{3}{4}}$ (d) $125^{\frac{2}{3}}$ (e) $16^{\frac{3}{2}}$

Answers:

(a) $9^{\frac{3}{2}} = (9^{\frac{1}{2}})^3 = (\sqrt{9})^3 = (3)^3 = 27$

(b) $32^{\frac{2}{5}} = (32^{\frac{1}{5}})^2 = (\sqrt[5]{32})^2 = (2)^2 = 4$

91

(5.2)

(c) $16^{\frac{3}{4}} = (16^{\frac{1}{4}})^3 = (\sqrt[4]{16}\,)^3 = (2)^3 = 8$

(d) $125^{\frac{2}{3}} = (125^{\frac{1}{3}})^2 = (\sqrt[3]{125}\,)^2 = (5)^2 = 25$

(e) $16^{\frac{3}{2}} = (16^{\frac{1}{2}})^3 = (\sqrt{16}\,)^3 = (4)^3 = 64$

VARIABLES AND RATIONAL EXPONENTS

If the base of the expression is a variable, use the Properties of Exponents to simplify the expression. (See Section 3.1 of this Study Guide.)

✠ *__Helpful Hint__* - For variables with rational exponents: (1) Do not use radicals. (2) Do not "split" the exponents. (3) Make sure that each final answer has positive exponents.

EXAMPLE 3   Simplify each expression:

(a) $x^{\frac{3}{2}}x^{\frac{1}{2}}$ (b) $\dfrac{z^{11/4}}{z^{3/4}}$ (c) $\left(x^{\frac{5}{3}}\right)^{\frac{9}{5}}$ (d) $\dfrac{y^{2/3}}{y^{3/4}}$ (e) $w^{\frac{3}{5}}w^2$

Answers:

(a) When multiplying, add the exponents of each variable. $x^{\frac{3}{2}}x^{\frac{1}{2}} = x^{\frac{3}{2}+\frac{1}{2}} = x^{\frac{4}{2}} = x^2$

(b) When dividing, subtract the exponents of each variable. $\dfrac{z^{11/4}}{z^{3/4}} = z^{\frac{11}{4}-\frac{3}{4}} = z^{\frac{8}{4}} = z^2$

(c) Multiply the outside exponent times the inside exponent. $\left(x^{\frac{5}{3}}\right)^{\frac{9}{5}} = x^{\left(\frac{5}{3}\right)\left(\frac{9}{5}\right)} = x^{\left(\frac{45}{15}\right)} = x^3$

(d) Subtract the exponents. $\dfrac{y^{2/3}}{y^{3/4}} = y^{\frac{2}{3}-\frac{3}{4}}$

Simplify the exponent using the lowest common denominator 12.

$$\frac{2}{3} - \frac{3}{4} = \frac{2}{3}\left(\frac{4}{4}\right) - \frac{3}{4}\left(\frac{3}{3}\right) = \frac{8}{12} - \frac{9}{12} = -\frac{1}{12}$$

Finish the original problem: $\dfrac{y^{2/3}}{y^{3/4}} = y^{\frac{2}{3}-\frac{3}{4}} = y^{-\frac{1}{12}} = \dfrac{1}{y^{1/12}}$

(e) Add the exponents. $w^{\frac{3}{5}}w^2 = w^{\frac{3}{5}+\frac{2}{1}} = w^{\frac{3}{5}+\frac{2}{1}\left(\frac{5}{5}\right)} = w^{\frac{3}{5}+\frac{10}{5}} = w^{\frac{13}{5}}$

5.3   SIMPLIFYING RADICALS

REVIEW OF REDUCING RADICALS   (Part 1 – The radicand reduces completely)

For the following properties, N is a positive integer and the radicals $\sqrt[N]{x}$ and $\sqrt[N]{y}$ are real numbers.

PROPERTY 1 – The "Nth" root cancels out the "Nth" power: $\boxed{\sqrt[N]{x^N} = x}$

Recall that if the radicand is a variable, divide the index into the exponent of the variable.

EXAMPLE 1   (a) $\sqrt{x^2} = x$ (b) $\sqrt[3]{y^3} = y$ (c) $\sqrt[8]{z^8} = z$

PROPERTY 2 – **Product Rule for Radicals**

If the radicand is a product, the "Nth" root is applied to each factor in the radicand:

$$\boxed{\sqrt[N]{xy} = \sqrt[N]{x}\ \sqrt[N]{y}}$$

✠ *Helpful Hint* – Refer to Section 5.1, Example 4. Use Property 2 to simplify a radical involving the product of a number and variable(s):
- Write the "Nth" root of the coefficient.
- Divide the index into the exponent of each variable.

EXAMPLE 2   Simplify: (a) $\sqrt{49x^2}$ (b) $\sqrt[3]{125x^3}$ (c) $\sqrt[3]{-8r^3s^6}$

   (d) $\sqrt[4]{81x^8}$ (e) $\sqrt[5]{32a^{15}b^5}$ (f) $\sqrt[5]{-100,000r^{20}s^{25}}$

Answers:

(a) $\sqrt{49x^2} = 7x$

(b) $\sqrt[3]{125x^3} = 5x$

(c) $\sqrt[3]{-8r^3s^6} = -2rs^2$

(d) $\sqrt[4]{81x^8} = 3x^2$

(e) $\sqrt[5]{32a^{15}b^5} = 2a^3b$

(f) $\sqrt[5]{-100,000r^{20}s^{25}} = -10r^4s^5$

PROPERTY 3 – **Quotient Rule for Radicals**

If the radicand is a quotient, the "Nth" root is applied to both the numerator and denominator:

$$\boxed{\sqrt[N]{\dfrac{x}{y}} = \dfrac{\sqrt[N]{x}}{\sqrt[N]{y}}} \quad \text{and} \quad \boxed{\dfrac{\sqrt[N]{x}}{\sqrt[N]{y}} = \sqrt[N]{\dfrac{x}{y}}}$$

EXAMPLE 3   (a) $\sqrt{\dfrac{4}{9}} = \dfrac{\sqrt{4}}{\sqrt{9}} = \dfrac{2}{3}$ (b) $\sqrt{\dfrac{25}{36}} = \dfrac{5}{6}$ (c) $\sqrt[3]{\dfrac{x^3}{8}} = \dfrac{x}{2}$

   (d) $\dfrac{\sqrt[3]{16}}{\sqrt[3]{2}} = \sqrt[3]{\dfrac{16}{2}} = \sqrt[3]{8} = 2$

(5.3)

SIMPLIFYING RADICALS (Part 2 – The radicand does not reduce completely)

For the radical $\sqrt{4}$, the radicand 4 is a perfect square. Therefore $\sqrt{4}$ can be reduced as $\sqrt{4} = 2$.
For the radical $\sqrt{45}$, the radicand 45 is not a perfect square. However, 9 is a perfect square which divides 45 evenly. We will rewrite $\sqrt{45}$ in simplified form by using the following property:

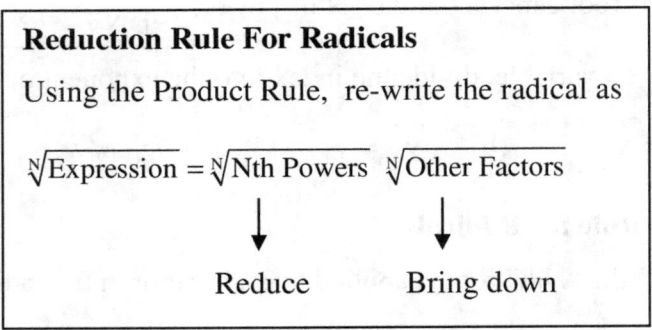

Reduction Rule For Radicals

Using the Product Rule, re-write the radical as

$$\sqrt[N]{\text{Expression}} = \sqrt[N]{\text{Nth Powers}}\ \sqrt[N]{\text{Other Factors}}$$

↓ Reduce ↓ Bring down

Numerical Radicals

Helpful Hint – First list the powers which are lower than the radicand. In the first radical after the equal sign, use the largest power in your list which divides the radicand evenly. (Refer to Step (a) in the examples below.)

EXAMPLE 4 Simplify $\sqrt{45}$.

Answer:

(a) For a square root, list the "perfect squares" lower than 45. The largest "square" which divides 45 evenly is **9**. 1, 4, **9**, 16, 25, 36, . . .

(b) Set up $\sqrt{45}$ as the product of two square roots. $\sqrt{45} = \sqrt{\ } \cdot \sqrt{\ }$
Since **9** $\cdot$ 5 = 45, place **9** in the first radical and place 5 in the second radical. $\sqrt{45} = \sqrt{9} \cdot \sqrt{5}$

(c) Reduce the first radical. Bring down the second radical. $= 3\sqrt{5}$

EXAMPLE 5 Simplify $\sqrt{32}$.

Answer:

(a) For the square root, list the "perfect squares" lower than 32. The largest "square" which divides 32 evenly is **16**. 1, 4, 9, **16**, 25, . . .

(b) Set up $\sqrt{32}$ as the product of two square roots. $\sqrt{32} = \sqrt{\ } \cdot \sqrt{\ }$
Since **16** $\cdot$ 2 = 32, place **16** in the first radical and place 2 in the second radical. $\sqrt{32} = \sqrt{16} \cdot \sqrt{2}$

(c) Reduce the first radical. Bring down the second radical. $= 4\sqrt{2}$

EXAMPLE 6 Simplify $\sqrt[3]{24}$.

Answer:

(a) For a cube root, list some of the "perfect cubes".
The largest "cube" which divides 24 evenly is **8**.

1, **<u>8</u>**, 27, 64, 125, . . .

(b) Set up $\sqrt[3]{24}$ as the product of two cube roots.

Since **8** · 3 = 24, place **8** in the first radical and place 3 in the second radical.

$\sqrt[3]{24} = \sqrt[3]{} \cdot \sqrt[3]{}$

$\sqrt[3]{24} = \sqrt[3]{8} \cdot \sqrt[3]{3}$

(c) Reduce the first radical. Bring down the second radical.

$= 2\sqrt[3]{3}$

EXAMPLE 7 Simplify $\sqrt[4]{48}$.

Answer:

(a) For a 4th root, list the some of the "4th powers"
The largest "4th power" which divides 48 evenly is **16**.

1, **<u>16</u>**, 81, 256 . . .

(b) Set up $\sqrt[4]{48}$ as the product of two 4th roots.

Since **16** · 3 = 48, place **16** in the first radical and place 3 in the second radical.

$\sqrt[4]{48} = \sqrt[4]{} \cdot \sqrt[4]{}$

$\sqrt[4]{48} = \sqrt[4]{16} \cdot \sqrt[4]{3}$

(c) Reduce the first radical. Bring down the second radical.

$= 2\sqrt[4]{3}$

Radicals Involving Variables

<u>Helpful Hints</u>:

- In the first radical after the equal sign: For the variable, use the *closest exponent* which the index divides into evenly.
- Place the other factor inside the second radical.
- Reduce the first radical by dividing the index into the exponent. Bring down the second radical. Each exponent in the second radical must be smaller than the index.

EXAMPLE 8 Simplify $\sqrt{x^7}$.

Answer:

(a) Set up the product of two square roots.

$\sqrt{x^7} = \sqrt{}\ \sqrt{}$

(b) For x^7, the index 2 does not divide into 7 evenly.
The exponent closest to 7 that works is the 6th power.
Place $\mathbf{x^6}$ in the first radical and x^1 in the second radical.

$= \sqrt{x^6}\ \sqrt{x^1}$

(c) Reduce the first radical by dividing the index 2 into the exponent 6. Bring down the second radical.

$= x^3\sqrt{x}$

(5.3)

EXAMPLE 9   Simplify $\sqrt[3]{x^7}$.

Answer:

(a) Set up the product of two cube roots.

$$\sqrt[3]{x^7} = \sqrt[3]{} \; \sqrt[3]{}$$

(b) For x^7, the index 3 does not divide into 7 evenly.
The exponent closest to 7 that works is the 6th power.

Place $\mathbf{x^6}$ in the first radical and x^1 in the second radical.

$$= \sqrt[3]{x^6} \; \sqrt[3]{x^1}$$

(c) Reduce the first radical. Bring down the second radical.

$$= x^2 \sqrt[3]{x}$$

EXAMPLE 10   Simplify $\sqrt[4]{x^7}$.

Answer:

(a) Set up the product of two 4th roots.

$$\sqrt[4]{x^7} = \sqrt[4]{} \; \sqrt[4]{}$$

(b) For x^7, the index 4 does not divide into 7 evenly.
The exponent closest to 7 that works is the 4th power.

Place $\mathbf{x^4}$ in the first radical and x^3 in the second radical.

$$= \sqrt[4]{x^4} \; \sqrt[4]{x^3}$$

(c) Reduce the first radical. Bring down the second radical.

$$= x \sqrt[4]{x^3}$$

EXAMPLE 11   Simplify $\sqrt[8]{x^7}$.
Answer:
For x^7, the index 8 does not divide into 7 evenly because the exponent 7 is too small.
Therefore $\sqrt[8]{x^7}$ cannot be reduced.

✖ *Helpful Hints:* **Radicals Involving A Coefficient And Variables**

- For the "nth" root, list some of the "nth" powers.
- In the first radical after the equal sign, write the largest power which divides into the coefficient evenly. Place the other numerical factor inside the second radical.
- In the first radical after the equal sign: For each variable, use the closest exponent which the index divides into evenly.
- Place the other variable factors inside the second radical.
- Reduce the first radical and bring down the second radical.
- Each exponent in the second radical must be smaller than the index.

EXAMPLE 12   Simplify $\sqrt{24x^3y^5}$.
Answer:

(a) List some of the "squares".   1, $\underline{\mathbf{4}}$, 9, 16, 25, . . .
The largest "square" which divides 24 evenly is **4**.

(b) Set up the product of two square roots.

$$\sqrt{24x^3y^5} = \sqrt{} \; \sqrt{}$$

96

(5.3)

Since 24 = 4 · 6, place **4** in the first radical.

$$\sqrt{24x^3y^5} = \sqrt{}\ \sqrt{}$$

Place 6 in the second radical.

$$\sqrt{4}\ \ \sqrt{6}$$

(c) For x^3, the index 2 does not divide into 3 evenly. The closest exponent that works is 2nd power.

Place x^2 in the first radical and x^1 in the second radical.

$$\sqrt{4x^2}\ \ \sqrt{6x^1}$$

(d) For y^5, the index 2 does not divide into 5 evenly. The closest exponent that works is the 4th power.

Place y^4 in the first radical and y^1 in the second radical.

$$\sqrt{4x^2y^4}\ \sqrt{6x^1y^1}$$

(g) Reduce the first radical and bring down the second radical.

$$\sqrt{24x^3y^5} = 2xy^2\sqrt{6xy}$$

EXAMPLE 13   Simplify $\sqrt[3]{54w^{16}x^{11}y^6z^2}$.

Answer:

(a) List some of the "cubes".
The largest "cube" which divides 54 evenly is **27**.

$$1,\ 8,\ \underline{\mathbf{27}},\ 64,\ \ldots$$

(b) Set up the product of two cube roots.

$$\sqrt[3]{54w^{16}x^{11}y^6z^2} =$$

Since 54 = 27 · 2, place **27** in the first radical.

$$\sqrt[3]{}\ \ \sqrt[3]{}$$

Place 2 in the second radical.

$$\sqrt[3]{27}\ \ \ \sqrt[3]{2}$$

(c) For w^{16}, the index 3 does not divide into 16 evenly. The closest exponent that works is 15th power.

Place w^{15} in the first radical and w^1 in the second radical.

$$\sqrt[3]{27w^{15}}\ \ \ \sqrt[3]{2w^1}$$

(d) For x^{11}, the index 3 does not divide into 11 evenly. The closest exponent that works is the 9th power.

Place x^9 in the first radical and x^2 in the second radical.

$$\sqrt[3]{27w^{15}x^9}\ \ \ \sqrt[3]{2w^1x^2}$$

(e) For y^6, the index 3 divides into 6 evenly.

Place y^6 in the first radical.

$$\sqrt[3]{27w^{15}x^9y^6}\ \sqrt[3]{2w^1x^2}$$

(f) For z^2, the index 3 does not divide into 2 evenly because this exponent is too small.

Place z^2 in the *second* radical.

$$\sqrt[3]{27w^{15}x^9y^6}\ \sqrt[3]{2w^1x^2z^2}$$

(g) Reduce the first radical.
Bring down the second radical.

$$\sqrt[3]{54w^{16}x^{11}y^6z^2} = 3w^5x^3y^2\sqrt[3]{2wx^2z^2}$$

5.4   ADDING AND SUBTRACTING RADICAL EXPRESSIONS

For radicals, **similar terms** have the same index and the same radicand. The expressions $5\sqrt{2}$, $4\sqrt{2}$ and $-3\sqrt{2}$ are similar terms. In general, **similar terms** have the same variables, with the same exponents, and the same radical.

EXAMPLE 1   For the following list of expressions, identify the similar terms:

$$7x^2\sqrt{2}, \ \sqrt{2}, \ -8x\sqrt[3]{2}, \ -5\sqrt{2}, \ 5x\sqrt[3]{2}, \ 3x\sqrt{2}, \ 3x^2\sqrt{2}, \ 4$$

Answer: $7x^2\sqrt{2}$ and $3x^2\sqrt{2}$ are similar terms.   (Same variable, same exponent, same radical)

$\sqrt{2}$ and $-5\sqrt{2}$ are similar terms.   (No variable and same radical)

$-8x\sqrt[3]{2}$ and $5x\sqrt[3]{2}$ are similar terms.   (Same variable, same exponent, same radical)

$3x\sqrt{2}$ is <u>not</u> similar to any of the terms in the list.   (Different exponent on x, or, different radical)

4 is <u>not</u> similar to any of the terms in the list.   (4 is not a radical.)

ADDITION RULE FOR RADICALS:

> To simplify the sum or difference of radicals, add or subtract the coefficients of similar terms.

EXAMPLE 2   Simplify: (a) $5\sqrt{2} + 4\sqrt{2} - 3\sqrt{2}$ (b) $7x^2\sqrt{2} - 8x\sqrt[3]{2} + 3x^2\sqrt{2} + 5x\sqrt[3]{2}$

Answers:

(a) $5\sqrt{2} + 4\sqrt{2} - 3\sqrt{2} = 6\sqrt{2}$

(b) $7x^2\sqrt{2} - 8x\sqrt[3]{2} + 3x^2\sqrt{2} + 5x\sqrt[3]{2} = 10x^2\sqrt{2} - 3x\sqrt[3]{2}$

✳ ***Helpful Hint*** – To simplify an expression such as $10xy\sqrt{45} - 3y\sqrt{20x^2} + 7\sqrt{5x^2y^2}$, first observe that the three terms are not similar terms in the given format. Also, each of the radicals can be reduced. (1) Rewrite each term in reduced radical form. (2) Then add or subtract the similar terms.

EXAMPLE 3   Simplify: $10xy\sqrt{45} - 3y\sqrt{20x^2} + 7\sqrt{5x^2y^2}$

Answer:

List some of the "squares".   Squares: 1, **4** , **9** , 16, 25, 36, 49, . . .

Rewrite each radical as the product of two square roots. $10xy\sqrt{45} - 3y\sqrt{20x^2} + 7\sqrt{5x^2y^2}$

$45 = \mathbf{9} \cdot 5$ and $20 = \mathbf{4} \cdot 5$ $= 10xy\sqrt{9}\sqrt{5} - 3y\sqrt{4x^2}\sqrt{5} + 7\sqrt{x^2y^2}\sqrt{5}$

Simplify each term. $= 10xy(3)\sqrt{5} - 3y(2x)\sqrt{5} + 7(xy)\sqrt{5}$

$= 30xy\sqrt{5} - 6xy\sqrt{5} + 7xy\sqrt{5}$

The three terms are now similar terms.

Simplify the expression. $= 31xy\sqrt{5}$

EXAMPLE 4   Simplify: $-12a \sqrt[3]{16a^2b^4} + 7b \sqrt[3]{54a^5b}$

Answer:

List the "cubes".   Cubes: 1, **8**, **27**, 64, 125, . . .

Rewrite each radical as the product of two cube roots. $-12a \sqrt[3]{16a^2b^4} + 7b \sqrt[3]{54a^5b}$

$16 = \mathbf{8} \cdot 2$ and $54 = \mathbf{27} \cdot 2$ $= -12a \sqrt[3]{8b^3} \sqrt[3]{2a^2b} + 7b \sqrt[3]{27a^3} \sqrt[3]{2a^2b}$

Simplify each term. $= -12a\,(2b) \sqrt[3]{2a^2b} + 7b\,(3a) \sqrt[3]{2a^2b}$

$= -24ab \sqrt[3]{2a^2b} + 21ab \sqrt[3]{2a^2b}$

Both terms are now similar terms.
Simplify the expression. $= -3ab \sqrt[3]{2a^2b}$

5.5   MULTIPLICATION AND DIVISION OF RADICALS

MULTIPLYING RADICALS

PRODUCT RULE FOR RADICALS $\boxed{\sqrt[N]{xy} = \sqrt[N]{x}\ \sqrt[N]{y} \quad \text{and} \quad \sqrt[N]{x}\ \sqrt[N]{y} = \sqrt[N]{xy}}$

Recall that we used the first part of the Product Rule to reduce a radical involving a product. We use the second part of the Product Rule to simplify the product of two radicals having the same index. Observe that we multiply the radicands. Write the final answer in reduced (simplified) form.

EXAMPLE 1   (a) $(\sqrt{2})(\sqrt{8}) = \sqrt{16} = 4$ (b) $\sqrt{3x}\sqrt{12x} = \sqrt{36x^2} = 6x$

(c) $\sqrt[3]{2x^2}\sqrt[3]{4x} = \sqrt[3]{8x^3} = 2x$ (d) $\sqrt{5}\sqrt{5} = \sqrt{25} = 5$ (e) $\sqrt{x}\sqrt{x} = \sqrt{x^2} = x$

✖ *__Helpful Hint__* – Concerning Square Roots:
If a square root is multiplied times itself, the square root symbol "cancels out": $\boxed{\sqrt{x}\sqrt{x} = x}$

(f) $\sqrt{y}\sqrt{y} = y$ (g) $\sqrt{2}\sqrt{2} = 2$ (h) $\sqrt{3y}\sqrt{3y} = 3y$

BASIC MULTIPLICATION RULES FOR RADICALS

> To multiply two radical expressions which have the same index:
> • Multiply the outside factors and multiply the inside factors.
> • Then reduce to simplified form.

EXAMPLE 2   (a) $(2\sqrt{3})(5\sqrt{12}) = 10\sqrt{36} = 10(6) = 60$

✖ *__Helpful Hint__* – When you multiply square roots, remember to check the radicand for division by 4, 9, or any other perfect square. Always reduce the answer to simplified form.

(b) $(\sqrt{21})(\sqrt{3}) = \sqrt{63} = \sqrt{9}\sqrt{7} = 3\sqrt{7}$

(5.5)

EXAMPLE 3   Simplify: $(3a)\left(4\sqrt{2b}\right)\left(5\sqrt{7}\right)$

Answer:

$(3a)\left(4\sqrt{2b}\right)\left(5\sqrt{7}\right)$

Multiply the outside factors and multiply the inside factors.

$= 60a\sqrt{14b}$

EXAMPLE 4   Simplify: $\sqrt[3]{4x^2}\ \sqrt[3]{4x}$

Answer:

$\sqrt[3]{4x^2}\ \sqrt[3]{4x}$

Multiply the inside factors.

$= \sqrt[3]{16x^3}$

Reduce the answer to simplified form using $16 = (8)(2)$.

$= \sqrt[3]{8x^3}\ \sqrt[3]{2}$

$= 2x\ \sqrt[3]{2}$

EXAMPLE 5   Simplify: $\left(5\sqrt[3]{6r^4s}\right)\left(2\sqrt[3]{9rs}\right)$

Answer:

$\left(5\sqrt[3]{6r^4s}\right)\left(2\sqrt[3]{9rs}\right)$

Multiply the outside factors and multiply the inside factors.

$= 10\sqrt[3]{54r^5s^2}$

Reduce the answer to simplified form using $54 = (27)(2)$.

$= 10\sqrt[3]{27r^3}\ \sqrt[3]{2r^2s^2}$

$= 10(3r)\ \sqrt[3]{2r^2s^2}$

$= 30r\ \sqrt[3]{2r^2s^2}$

EXAMPLE 6   Simplify: $\left(\sqrt{15bc}\right)\left(\sqrt{2ab^5}\right)\left(\sqrt{3ab}\right)$

Answer:

$\left(\sqrt{15bc}\right)\left(\sqrt{2ab^5}\right)\left(\sqrt{3ab}\right)$

Multiply the inside factors.

$= \sqrt{90a^2b^7c}$

Reduce the answer to simplified form using $90 = (9)(10)$.

$\sqrt{9a^2b^6}\ \sqrt{10bc}\ = 3ab^3\sqrt{10bc}$

BINOMIAL PRODUCTS AND SQUARE ROOTS

For each of the following examples, use the Algebra rules for multiplying binomials.

EXAMPLE 7   Simplify: $2\sqrt{7}\,(4-3\sqrt{7}\,)$

Answer: Use the Distributive Property to multiply $2\sqrt{7}$ times each term of the binomial.

$2\sqrt{7}\,(4+3\sqrt{7}\,) = (2\sqrt{7})(4) + (2\sqrt{7})(3\sqrt{7}\,) = 8\sqrt{7} + (6)(\sqrt{49}) = 8\sqrt{7} + (6)(7)$

$= 42 + 8\sqrt{7}$

EXAMPLE 8   Simplify: $(3\sqrt{2} + \sqrt{5})(4\sqrt{2} - \sqrt{5}\,)$

Answer: Use the F.O.I.L. Method to multiply two binomial forms.

$(3\sqrt{2} + \sqrt{5})(4\sqrt{2} - \sqrt{5}\,) = (3\sqrt{2})(4\sqrt{2}) + (3\sqrt{2})(-\sqrt{5}) + (\sqrt{5})(4\sqrt{2}) + (\sqrt{5})(-\sqrt{5}\,)$

$= 12\sqrt{4} - 3\sqrt{10} + 4\sqrt{10} - \sqrt{25} = 12(2) + \sqrt{10} - 5 = 24 + \sqrt{10} - 5 = 19 + \sqrt{10}$

EXAMPLE 9 Simplify: $(2\sqrt{5}-3)^2$

Answer: Rewrite the problem as the binomial times itself, (Binomial)(Binomial). Then use the F.O.I.L. Method to simplify the expression.

$$\begin{aligned}(2\sqrt{5}-3)^2 &= (2\sqrt{5}-3)(2\sqrt{5}-3)\\ &= (2\sqrt{5})(2\sqrt{5})+(2\sqrt{5})(-3)+(-3)(2\sqrt{5})+(-3)(-3)\\ &= (4)(\sqrt{25})+(-6)(\sqrt{5})+(-6)(\sqrt{5})+9\\ &= (4)(5) - 6\sqrt{5} - 6\sqrt{5}+9 = 20 - 12\sqrt{5}+9\\ &= 29 - 12\sqrt{5}\end{aligned}$$

EXAMPLE 10 Simplify: $(\sqrt{11}+\sqrt{3})(\sqrt{11}-\sqrt{3})$

Answer: This problem is the **product of conjugates**. Recall that conjugates are binomials which have the same terms but different signs. Use the shortcut for multiplying conjugates:

Write the square of the 1st term minus the square of the 2nd term.

$$\begin{aligned}(\sqrt{11}+\sqrt{3})(\sqrt{11}-\sqrt{3}) &= (\sqrt{11})^2 - (\sqrt{3})^2\\ &= 11-3 = 8\end{aligned}$$

(Note: You may also work the product of conjugates by using F.O.I.L.)

DIVISION AND RADICALS

DIVISION RULE 1 – **A Radical Containing a Fraction:**

When you simplify an expression involving radicals, the radicand should not contain a fraction. Use the Quotient Rule to separate the radical and reduce the denominator.

EXAMPLE 11 $\sqrt{\dfrac{5}{16}} = \dfrac{\sqrt{5}}{\sqrt{16}} = \dfrac{\sqrt{5}}{4}$

DIVISION RULE 2 – **A Radical in the Denominator of a Fraction:**

When you simplify an expression involving radicals, the denominator of a fraction should not contain a radical. Rationalize the denominator.

Helpful Hint – "Rationalize the denominator" means rewrite the denominator without using a radical symbol. In other words, write the denominator in a format which will reduce the radical.

EXAMPLE 12 (a) $\dfrac{3}{\sqrt{25}} = \dfrac{3}{5}$ (b) $\sqrt{98} \div \sqrt{72} = \dfrac{\sqrt{98}}{\sqrt{72}} = \sqrt{\dfrac{98}{72}} = \sqrt{\dfrac{49}{36}} = \dfrac{7}{6}$

(5.5)

Monomial Denominator Containing a Square Root

If a square root in the denominator cannot be reduced, *rationalize the denominator:* Multiply the fraction times the unit fraction formed by the denominator over itself. Then simplify the expression:

$$\frac{\text{Numerator}}{\sqrt{X}} \bullet \left[\frac{\sqrt{X}}{\sqrt{X}}\right] = \frac{\text{New Numerator}}{X}$$

EXAMPLE 13 Rationalize the denominator: (a) $\frac{2}{\sqrt{7}}$ (b) $\sqrt{\frac{2}{5}}$

Answers:

(a) $\frac{2}{\sqrt{7}} = \frac{2}{\sqrt{7}} \cdot \left[\frac{\sqrt{7}}{\sqrt{7}}\right] = \frac{2\sqrt{7}}{7}$

(b) Apply the Quotient Rule to separate the square root. $\sqrt{\frac{2}{5}} = \frac{\sqrt{2}}{\sqrt{5}}$

Rationalize the square root in the denominator. $= \frac{\sqrt{2}}{\sqrt{5}} \cdot \left[\frac{\sqrt{5}}{\sqrt{5}}\right]$

Simplify the numerator and the denominator. $= \frac{\sqrt{10}}{5}$

Helpful Hint – In a fraction, *a term outside the radical never cancels a term inside the radical.*

Therefore $\frac{\sqrt{10}}{5}$ cannot be reduced.

EXAMPLE 14 Rationalize the denominator: (a) $\frac{4}{\sqrt{5x}}$ (b) $\sqrt{\frac{3}{8}}$

Answer:

(a) Multiply $\frac{4}{\sqrt{5x}}$ by the unit fraction $\frac{\sqrt{5x}}{\sqrt{5x}}$. $\frac{4}{\sqrt{5x}} = \frac{4}{\sqrt{5x}} \cdot \frac{\sqrt{5x}}{\sqrt{5x}}$

Simplify the numerator and the denominator. $= \frac{4\sqrt{5x}}{5x}$

(b) Use the Quotient Rule to separate the square root. $\sqrt{\frac{3}{8}} = \frac{\sqrt{3}}{\sqrt{8}}$

Reduce the denominator to simplified form using 8 = (4)(2). $= \frac{\sqrt{3}}{\sqrt{4}\sqrt{2}} = \frac{\sqrt{3}}{2\sqrt{2}}$

Multiply the fraction times $\frac{\sqrt{2}}{\sqrt{2}}$. $= \frac{\sqrt{3}}{2\sqrt{2}} \cdot \left[\frac{\sqrt{2}}{\sqrt{2}}\right] = \frac{\sqrt{6}}{2 \cdot 2}$

Simplify the denominator. $= \frac{\sqrt{6}}{4}$

Monomial Denominator Containing Other Roots

If the radical in the denominator has an index higher than square root and exponent A (inside the radicand) is too small, use the following procedure to *rationalize the denominator:*

$$\frac{\text{Numerator}}{\sqrt[N]{X^A}} \cdot \left[\frac{\sqrt[N]{X^B}}{\sqrt[N]{X^B}}\right] = \frac{\text{New Numerator}}{\sqrt[N]{X^N}} = \frac{\text{New Numerator}}{X}$$

Increase the exponent of the denominator. Fill in exponent B so that the sum of the exponents A + B = Index N.

EXAMPLE 15   Rationalize the denominator: (a) $\dfrac{2}{\sqrt[3]{x}}$ (b) $\sqrt[3]{\dfrac{125}{2}}$ (c) $\dfrac{3}{\sqrt[4]{a^2b^3}}$ (d) $\sqrt[5]{\dfrac{3}{2r^2}}$

(a) Given $x^1 \cdot \mathbf{x^2} = x^3$, multiply $\dfrac{2}{\sqrt[3]{x^1}}$ times $\dfrac{\sqrt[3]{x^2}}{\sqrt[3]{x^2}}$.

$$\frac{2}{\sqrt[3]{x}} = \frac{2}{\sqrt[3]{x^1}} \cdot \left[\frac{\sqrt[3]{x^2}}{\sqrt[3]{x^2}}\right] = \frac{2\sqrt[3]{x^2}}{\sqrt[3]{x^3}} = \frac{2\sqrt[3]{x^2}}{x}$$

(b) Use the Quotient Rule to separate the cube root.

$$\sqrt[3]{\frac{125}{2}} = \frac{\sqrt[3]{125}}{\sqrt[3]{2}} = \frac{5}{\sqrt[3]{2^1}}.$$

Given $2^1 \cdot \mathbf{2^2} = 2^3$, multiply $\dfrac{5}{\sqrt[3]{2^1}}$ times $\dfrac{\sqrt[3]{2^2}}{\sqrt[3]{2^2}}$.

$$\frac{5}{\sqrt[3]{2^1}} \cdot \left[\frac{\sqrt[3]{2^2}}{\sqrt[3]{2^2}}\right] = \frac{5\sqrt[3]{2^2}}{\sqrt[3]{2^3}} = \frac{5\sqrt[3]{4}}{2}$$

(c) Given $a^2b^3 \cdot \mathbf{a^2b^1} = a^4b^4$, multiply $\dfrac{3}{\sqrt[4]{a^2b^3}}$ times $\dfrac{\sqrt[4]{a^2b^1}}{\sqrt[4]{a^2b^1}}$:

$$\frac{3}{\sqrt[4]{a^2b^3}} \cdot \left[\frac{\sqrt[4]{a^2b^1}}{\sqrt[4]{a^2b^1}}\right] = \frac{3\sqrt[4]{a^2b^1}}{\sqrt[4]{a^4b^4}} = \frac{3\sqrt[4]{a^2b}}{ab}$$

(d) Use the Quotient Rule to separate the 5th root.

$$\sqrt[5]{\frac{3}{2r^2}} = \frac{\sqrt[5]{3}}{\sqrt[5]{2^1r^2}}.$$

Given $2^1r^2 \cdot \mathbf{2^4r^3} = 2^5r^5$, multiply $\dfrac{\sqrt[5]{3}}{\sqrt[5]{2^1r^2}}$ times $\dfrac{\sqrt[5]{2^4r^3}}{\sqrt[5]{2^4r^3}}$:

$$\frac{\sqrt[5]{3}}{\sqrt[5]{2^1r^2}} \cdot \left[\frac{\sqrt[5]{2^4r^3}}{\sqrt[5]{2^4r^3}}\right] = \frac{\sqrt[5]{3 \cdot 16r^3}}{\sqrt[5]{2^5r^5}} = \frac{\sqrt[5]{48r^3}}{2r}$$

Binomial Denominators With Square Roots

Multiply the fraction by the conjugate of the denominator over itself:

$$\frac{\text{Numerator}}{\text{Denominator}} \cdot \left[\frac{\text{Conjugate of Denominator}}{\text{Conjugate of Denominator}}\right] = \frac{\text{New Numerator}}{\text{Square} - \text{Square}}$$

EXAMPLE 16   Rationalize the denominator: (Use F.O.I.L. to multiply across.)

$$\frac{\sqrt{7}+4}{\sqrt{7}-2} = \frac{\sqrt{7}+4}{\sqrt{7}-2} \cdot \left[\frac{\sqrt{7}+2}{\sqrt{7}+2}\right] = \frac{7+2\sqrt{7}+4\sqrt{7}+8}{7-4} = \frac{15+6\sqrt{7}}{3} = \frac{15}{3} + \frac{6\sqrt{7}}{3} = 5+2\sqrt{7}$$

EXAMPLE 17   Rationalize the denominator: (Use F.O.I.L. to multiply across.)

$$\frac{\sqrt{x}-2\sqrt{y}}{2\sqrt{x}+\sqrt{y}} = \frac{\sqrt{x}-2\sqrt{y}}{2\sqrt{x}+\sqrt{y}} \cdot \frac{2\sqrt{x}-\sqrt{y}}{2\sqrt{x}-\sqrt{y}} = \frac{2(x)-\sqrt{xy}-4\sqrt{xy}+2(y)}{4(x)-y} = \frac{2x-5\sqrt{xy}+2y}{4x-y}$$

5.6 EQUATIONS WITH RADICALS

Observe the illustration below:

Let x equal a constant, such as 3. $\mathbf{x = 3}$

Square both sides of the equation. $(x)^2 = (3)^2$

$x^2 = 9$

Now solve for x by using the Factoring Method. $x^2 - 9 = 0$

$(x - 3)(x + 3) = 0$

The correct answer is $\mathbf{x = 3}$, which matches the original equation statement.

But $x = -3$ *is not a correct answer because it does not match the original equation statement.*

Helpful Hint – If you use an exponent on both sides of an equation, *check the answers in writing!* As shown in the illustration above, some results may *not* be true for the original equation.

CANCELATION RULE FOR RADICALS

For the following property, the index N is a positive integer and the radical $\sqrt[N]{x}$ is a real number.

$$\sqrt[N]{x^N} = x \quad \text{and} \quad \left(\sqrt[N]{x}\right)^N = x$$

The first part of the Cancelation Rule states that the "Nth" root cancels out the "Nth" power. We used this property in previous sections of this chapter to reduce radicals.

The second part of the Cancelation Rule states that the "Nth" power cancels out the "Nth" root.

EXAMPLE 1 (a) $\left(\sqrt{x}\right)^2 = x$ (b) $\left(\sqrt[3]{y+2}\right)^3 = y + 2$

(c) $\left(\sqrt[4]{w^2+8w+1}\right)^4 = w^2 + 8w + 1$ (d) $\left(\sqrt[5]{v-3}\right)^5 = v - 3$

SOLVING RADICAL EQUATIONS

Steps:

- Isolate the radical: The radical must be on one side of the equation.
- To cancel out a square root, square both sides of the equation. To cancel out a cube root, cube both sides of the equation. To cancel out a 4th root, apply the 4th power to both sides of the equation. To cancel out a 5th root, apply the 5th power to both sides of the equation, etc.
- Use the Cancelation Rule for Radicals to simplify the equation.
- Solve the resulting equation.
- *Check the result(s) in writing.* (Some results may not be true for the original equation.)

(5.6)

EXAMPLE 2   Solve the equation: $\sqrt{2x+1} - 5 = 0$

Isolate the radical. (The radical must be on one side of the equation.)

$$\sqrt{2x+1} - 5 = 0$$
$$\phantom{\sqrt{2x+1}} + 5 + 5$$
$$\sqrt{2x+1} = 5$$

To cancel out a square root, square both sides of the equation. $\left(\sqrt{2x+1}\right)^2 = (5)^2$

Simplify the equation. $2x + 1 = 25$

Solve the equation for x. $2x = 24$

Possible solution: $x = 12$

Check the result in writing: Substitute 12 into the original equation:

$$\sqrt{2(12)+1} - 5 = \sqrt{25} - 5 = 5 - 5 = 0$$

Correct Answer: $x = 12$.

EXAMPLE 3   Solve the equation: $\sqrt[3]{5x+7} + 1 = -1$

Isolate the radical. (The radical must be on one side of the equation.)

$$\sqrt[3]{5x+7} + 1 = -1$$
$$\phantom{\sqrt[3]{5x+7}} - 1 -1$$
$$\sqrt[3]{5x+7} = -2$$

To cancel out the cube root, cube both sides of the equation. $\left(\sqrt[3]{5x+7}\right)^3 = (-2)^3$

Simplify the equation. $5x + 7 = -8$

Solve the equation for x. $5x = -15$

Possible solution: $x = -3$

Check the result in writing: Substitute -3 into the original equation:

$$\sqrt[3]{5(-3)+7} + 1 = \sqrt[3]{-15+7} + 1 = \sqrt[3]{-8} + 1 = -2 + 1 = -1$$

Correct Answer: $x = -3$.

(5.6)

EXAMPLE 4   Solve the equation: $\sqrt[4]{5x-9} + \sqrt[4]{3x+1} = 0$

Isolate one of the radicals.

$$\sqrt[4]{5x-9} + \sqrt[4]{3x+1} = 0$$
$$-\sqrt[4]{3x+1} \qquad -\sqrt[4]{3x+1}$$

(One radical must be on one side of the equation.)

$$\sqrt[4]{5x-9} = -\sqrt[4]{3x+1}$$

To cancel out the 4th root, apply the 4th power to both sides
of the equation.

$$\left(\sqrt[4]{5x-9}\right)^4 = \left(-\sqrt[4]{3x+1}\right)^4$$

Simplify the equation.

$$5x - 9 = 3x + 1$$

Solve the equation for x.

$$2x - 9 = 1$$
$$2x = 10$$

Possible solution: $x = 5$

Check the result in writing. Substitute 5 into the original equation:

$$\sqrt[4]{5(5)-9} + \sqrt[4]{3(5)+1} = 0$$
$$\sqrt[4]{16} + \sqrt[4]{16} = 0$$
$$2 + 2 = 0 \text{ is False!}$$

Correct Answer: No solution

EXAMPLE 5   Solve the equation: $x - 1 = \sqrt{8x+1}$

The radical is already isolated (by itself).

$$x - 1 = \sqrt{8x+1}$$

To cancel out the square root, square both sides. The binomial
must be in parentheses.

$$(x-1)^2 = \left(\sqrt{8x+1}\right)^2$$

Simplify the equation:
On the left side of the equation, write the binomial twice.
On the right side of the equation, the square cancels out
the square root symbol. Therefore bring down the radicand.
Then use the F.O.I.L. Method to simplify the left side.

$$(x - 1)(x - 1) = 8x + 1$$

$$x^2 - 1x - 1x + 1 = 8x + 1$$

$$x^2 - 2x + 1 = 8x + 1$$

The equation has an exponent; the equation must equal zero.
Cancel out the right side of the equation.
Factor the polynomial using the Greatest Common Factor Method.
Solve the equation for x.

$$-8x - 1 \qquad -8x - 1$$
$$x^2 - 10x = 0$$
$$x(x - 10) = 0$$
$$x = 0 \text{ or } x - 10 = 0$$

Possible solutions: $x = 0$ or $x = 10$

Check each result in writing:

Substitute 0 into the original equation:

$$x - 1 = \sqrt{8x + 1}$$

$$0 - 1 = \sqrt{8(0) + 1}$$

$$0 - 1 = \sqrt{0 + 1}$$

$$-1 = \sqrt{1}$$

This statement is *false* because $\sqrt{1} = 1$.

Correct Answer: x = 10.

Substitute 10 into the original equation:

$$x - 1 = \sqrt{8x + 1}$$

$$10 - 1 = \sqrt{8(10) + 1}$$

$$9 = \sqrt{80 + 1}$$

$$9 = \sqrt{81}$$

This statement is true.

EXAMPLE 6   **Application**- -Write the equation and solve:

The larger of two numbers is 10 more than 3 times the smaller number. If the smaller number is equal to the square root of the larger number, find the numbers.

Smaller number   x

Larger number   3x + 10

Write the equation: $x = \sqrt{3x + 10}$

The radical is already isolated (by itself).

$$x = \sqrt{3x + 10}$$

To cancel out the square root, square both sides.

$$(x)^2 = \left(\sqrt{3x + 10}\right)^2$$

Simplify the equation.

The equation has an exponent; the equation must equal zero.

$$x^2 = 3x + 10$$
$$-3x - 10 \qquad -3x - 10$$

Cancel out the right side of the equation.

$$x^2 - 3x - 10 = 0$$

Factor the polynomial using the Trinomial Method.

$$(x + 2)(x - 5) = 0$$

Solve the equation for *x*.

$$x + 2 = 0 \quad \text{or} \quad x - 5 = 0$$

Possible solutions: $x = -2$ or $x = 5$

Check each result in writing:

Substitute – 2 into the original equation:

$$-2 = \sqrt{3(-2) + 10}$$

$$-2 = \sqrt{-6 + 10}$$

$$-2 = \sqrt{4}$$

This statement is *false* because $\sqrt{4} = 2$.

Substitute 5 into the original equation:

$$5 = \sqrt{3(5) + 10}$$

$$5 = \sqrt{15 + 10}$$

$$5 = \sqrt{25}$$

This statement is true. Therefore x = 5.

Correct Answers: x = 5 (Smaller number)

3x + 10 = 3(5) + 10 = 25 (Larger number)

5.7   COMPLEX NUMBERS

REVIEW OF REAL SQUARE ROOTS

Given the expression $\sqrt{x} = \sqrt[2]{x}$, "2" is the index and "x" is the radicand. A square root is a real number if the radicand is positive or 0. Thus:

$$\sqrt{0} = 0, \text{ a real number} \qquad\qquad \sqrt{1} = 1, \text{ a real number}$$
$$\sqrt{49} = 7, \text{ a real number.} \qquad\qquad -\sqrt{49} = -7, \text{ a real number}$$

IMAGINARY NUMBERS

The square root of a negative number is called an **imaginary number**. Thus $\sqrt{-25}$ is an imaginary number.

RULE 1 – **Basic Rules for Imaginary Numbers**

The square root of a negative number must be rewritten as i times the square root of the positive number. The symbol i is called the **imaginary unit.**

> **Imaginary Number:** Let a be a positive real number.
> $$\sqrt{-1} = i \qquad \text{and} \qquad \sqrt{-a} = i\sqrt{a}$$

EXAMPLE 1   (a) $\sqrt{-7} = i\sqrt{7}$ (b) $\sqrt{-2} = i\sqrt{2}$ (c) $\sqrt{-25} = i\sqrt{25} = i \cdot 5 = 5i$

�֍ *Helpful Hints* – (1) To rewrite an imaginary number: If the square root cannot be reduced, place the symbol i in front of the square root, as shown in Example 1. (2) If the square root can be reduced, apply the *"Shortcut Notation"*: Place the symbol i after the reduced number, as shown in the following examples.

EXAMPLE 2   Simplify each imaginary number: (a) $\sqrt{-9}$ (b) $\sqrt{-100}$ (c) $\sqrt{-144}$

Answers:

(a) Since $\sqrt{9} = 3$ and $\sqrt{-1} = i$, $\sqrt{-9} = 3i$.   (b) $\sqrt{-100} = 10i$ (c) $\sqrt{-144} = 12i$

RULE 2 – **Reduced Form for Imaginary Numbers**

Remember that the square root of a negative number must be rewritten in reduced (simplified) form.

EXAMPLE 3   Simplify each imaginary number: (a) $\sqrt{-18}$ (b) $\sqrt{-24}$

Answers:

(a) Rewrite $\sqrt{-18}$ in imaginary form. $\sqrt{-18} = i\sqrt{18}$

Using $18 = (9)(2)$, reduce the answer to simplified form. $= i\sqrt{9}\sqrt{2}$

                                                       $= 3i\sqrt{2}$

(b) Rewrite $\sqrt{-24}$ in imaginary form. $\sqrt{-24} = i\sqrt{24}$

Using $24 = (4)(6)$, reduce the answer to simplified form. $= i\sqrt{4}\sqrt{6} = 2i\sqrt{6}$

COMPLEX NUMBERS

A real number is a monomial; it does not use the imaginary unit *i*. An imaginary number may be in monomial or binomial format; the expression contains an *i* term.

Type of Number	Form (*a* and *b* are real numbers)	Examples
Real Number	a (No "*i*" Term)	$7;\ -\dfrac{3}{5}\ ;\ 2.3$
Imaginary Number	bi or $a + bi$	$-7i\ ;\ 2 + 3i\ ;\ 4 - 5i$

A **complex number** is any real or imaginary number. The **Set of Complex Numbers** consists of all of the real numbers combined with all of the imaginary numbers.

$$\{\text{Complex Numbers}\} \ = \ \{\text{Real numbers}\} \ \cup \ \{\text{Imaginary Numbers}\}$$
$$\text{``a''} \qquad\qquad \text{``bi'' } or \text{ ``a + bi''}$$

All of the numbers used above, $7,\ -\dfrac{3}{5},\ 2.3,\ -7i,\ 2 + 3i$ and $4 - 5i$, are examples of complex numbers.

SIMPLIFYING COMPLEX NUMBERS

As you simplify complex numbers, write the imaginary answers in the format ***a + bi*** or ***a − bi***. (Write the real term first and the imaginary term second.)

Addition and Subtraction

Add or subtract the real terms; add or subtract the imaginary terms.

EXAMPLE 4 $(4 + 5i) - (-8 + 2i) = 4 + 5i + 8 - 2i = 12 + 3i$

Multiplication

RULE 3 – **Exponent Rule for Imaginary Numbers** $\boxed{\text{Given } \sqrt{-1} = i\ ,\ i^2 = -1}$

When multiplying imaginary numbers, i^2 must be changed to −1. Then simplify the expression.

EXAMPLE 5   Simplify: (a) $3i(-3 + 4i)$ (b) $(2 + 3i)(1 - i)$

Answers:

(a) $3i(-3 + 4i) = -9i + 12\mathbf{i}^2 = -9i + 12(-1) = -12 - 9i$

(b) $(3 + 2i)(7 - i) = 21 - 3i + 14i - 2i^2 = 21 + 11i - 2(-1) = 21 + 11i + 2 = 23 + 11i$

(5.7)

Imaginary Conjugates

Let *a* and *b* be two real numbers. Real conjugates are two binomials which have the same terms but different signs, such as a + b and a – b. To write the conjugate of an imaginary number, change the sign of the imaginary term:

Imaginary Number	Imaginary Conjugate
$2 + 3i$	$2 - 3i$
$4 - 7i$	$4 + 7i$
$-3i$	$3i$
i	$-i$

EXAMPLE 6 Multiplying Imaginary Conjugates:

$$(4-7i)(4+7i) = (4)^2 - (7i)^2 = 16 - 49i^2 = 16 - 49(-1) = 16 + 49 = 45 \text{ (a real number)}$$

Helpful Hint – In the example above, observe that *the product of two imaginary conjugates results in a real number.*

Division (Simplifying Fractions)

Recall that the imaginary unit *i* represents a radical, $\sqrt{-1}$. Therefore *i* cannot remain in the denominator of a fraction. If the denominator of a fraction is an imaginary number, rationalize the denominator by multiplying the fraction times the conjugate of the denominator over itself:

$$\frac{\text{Numerator}}{\text{Imaginary Number}} \bullet \left[\frac{\text{Conjugate of Denominator}}{\text{Conjugate of Denominator}}\right] = \frac{\text{New Numerator}}{\text{Real Number}}$$

EXAMPLE 7 Rationalize the denominator: (a) $\frac{8+3i}{i}$ (b) $\frac{2-\sqrt{-25}}{3+\sqrt{-4}}$

Answers:

(a) $$\frac{8+3i}{i} = \frac{8+3i}{i} \bullet \left[\frac{-i}{-i}\right] = \frac{-8i-3i^2}{-i^2} = \frac{-8i-3(-1)}{-(-1)} = \frac{-8i+3}{1} = 3-8i$$

(b) $$\frac{2-\sqrt{-25}}{3+\sqrt{-4}} = \frac{2-5i}{3+2i} = \frac{2-5i}{3+2i} \bullet \left[\frac{3-2i}{3-2i}\right] = \frac{6-4i-15i+10i^2}{9-4i^2}$$

$$= \frac{6-19i+10(-1)}{9-4(-1)} = \frac{6-19i-10}{9+4} = \frac{-4-19i}{13} = -\frac{4}{13} - \frac{19i}{13}$$

Chapter 6   Polynomial Equations and Inequalities

6.1   SOLVING QUADRATIC EQUATIONS

In Chapter 3, Sections 3.4-3.5, we studied the **quadratic equation**, $Ax^2 + Bx + C = 0$, where A, B and C are real numbers. We observed that a quadratic equation is a second-degree equation and has exactly 2 roots or solutions. We used the Factoring Method to solve the quadratic equation. In this chapter we will present four methods for solving quadratic equations:
   Factoring, Square Root, Complete-the-Square, and the Quadratic Formula

REVIEW OF THE FACTORING METHOD

EXAMPLE 1   Solve by using the Factoring Method: $x(x + 3) = 3x + 4$

The equation must equal 0. $\hspace{6cm} x(x + 3) = 3x + 4$

   Multiply the left side of the equation. $\hspace{4cm} x^2 + 3x = 3x + 4$

   Cancel the right side of the equation. $\hspace{5cm} x^2 - 4 = 0$

Factor the polynomial using the Difference of Squares method. $\hspace{1cm} (x - 2)(x + 2) = 0$

Solve for x. $\hspace{7cm} x = 2 \ \text{ or } \ x = -2$

SQUARE ROOT METHOD

If the quadratic equation has an x^2 term and constants but no monomial x terms, we use the Square Root Method to solve the equation.

EXAMPLE 2   Solve by using the square-root method: $x^2 - 8 = 0$

Solve for x^2. (Add 8 to both sides of the equation.) $\hspace{3cm} x^2 - 8 = 0$

$$x^2 = 8$$

To cancel out the "square", write the square root of both sides of the equation. $\hspace{1cm} \sqrt{x^2} = \pm\sqrt{8}$
   Remember that a second degree equation has two roots. Use "double signs"
   for the constant on the right side of the equation.

Simplify both sides of the equation. $\hspace{5cm} x = \pm\sqrt{4}\sqrt{2}$

Answers: $\hspace{2cm} x = \pm 2\sqrt{2}$

✠ ***Helpful Hint*** – In the example above, the statement $x = \pm 2\sqrt{2}$ represents two roots:
$x = 2\sqrt{2} \text{ or } x = -2\sqrt{2}$.

(6.1)

EXAMPLE 3   Solve by using the square-root method: $3x^2 - 5 = 0$

Solve for x^2 by adding 5 to both sides of the equation and dividing both sides of the equation by 3.

$$3x^2 - 5 = 0$$
$$3x^2 = 5$$
$$x^2 = \frac{5}{3}$$

To cancel out the "square", write the square root of both sides of the equation.

$$\sqrt{x^2} = \pm\sqrt{\frac{5}{3}}$$

Simplify both sides of the equation. Rationalize the denominator.

$$x = \pm\frac{\sqrt{5}}{\sqrt{3}}\left(\frac{\sqrt{3}}{\sqrt{3}}\right)$$

Answers: $x = \pm\dfrac{\sqrt{15}}{3}$

EXAMPLE 4   Solve by using the square-root method: $(x + 2)^2 = -45$

Write the square root of both sides of the equation.

$$(x + 2)^2 = -45$$

Use "double signs" for the constant on the right side of the equation.
Simplify both sides of the equation.

$$\sqrt{(x+2)^2} = \pm\sqrt{-45}$$
$$x + 2 = \pm i\sqrt{9}\sqrt{5}$$
$$x + 2 = \pm 3i\sqrt{5}$$

Subtract 2 on both sides of the equation.

$$\begin{array}{r} -2 \quad -2 \end{array}$$

Answers: $x = -2 \pm 3i\sqrt{5}$

COMPLETE-THE-SQUARE METHOD

In chapter 3 we learned to use F.O.I.L. to square a binomial, such as:

$$(a+3)^2 = (a+3)(a+3) = a^2 + 3a + 3a + 9 = a^2 + 6a + 9$$

The answer, $a^2 + 6a + 9$, is called a **perfect square trinomial**.

For the Complete-the-Square Method, we are given the first two terms of a trinomial. The objective is to fill in the third term and create a perfect square trinomial which will factor as a (binomial)^2:

$$a^2 + 6a + \underline{} \quad \rightarrow \quad a^2 + 6a + \underline{9} \quad \rightarrow \quad (a+3)^2$$

Steps for Completing the Square

- Write the first two terms of the trinomial plus a blank space.
- To determine the third term, multiply $\dfrac{1}{2}$ times coefficient of the second term and square the result.
- Write this number in the blank space.
- Then rewrite the trinomial as a (binomial)^2.

(6.1)

The following practice problems in Example 5 illustrate the steps for completing the square.
(In Example 6 we will use the Complete-the-Square Method to solve a quadratic equation.)

EXAMPLE 5   For each expression, *complete the square* by adding the third term required to create a perfect square trinomial. Rewrite the trinomial as the square of a binomial.

(a) $y^2 + 10y + \underline{\,?\,}$ 　　　　　　(b) $x^2 - 4x + \underline{\,?\,}$ 　　　　　　(c) $r^2 + 5r + \underline{\,?\,}$

Answers:

(a) The coefficient of the second term is 10. To Complete the Square: 　　　　$y^2 + 10y + \underline{\,?\,}$

Multiply $\frac{1}{2} \cdot$ coefficient of the second term and square the result. 　　　　$\frac{1}{2} \cdot 10 = 5$ and $5^2 = \underline{\mathbf{25}}$

　　(5 is the "half number" and 25 is the "square number")

Write the "square number" 25 in the blank space. 　　　　　　$y^2 + 10y + \underline{\,\mathbf{25}\,}$

The "half number" 5 is positive. Rewrite the trinomial as the square 　　　$= (y + \mathbf{5})^2$
　of the binomial y + 5.

❉ *Helpful Hint* - Refer to Example 5(a) above. **Shortcut for the Complete-the-Square Method:**
　　The "square number" goes to the trinomial and the "half number" goes to the binomial.

In other words, the "square number" is the third term of the trinomial and the "half number" is the second term of the binomial.

(b) The coefficient of the second term is -4. To Complete the Square: 　　　$x^2 - 4x + \underline{\,?\,}$

Multiply $\frac{1}{2} \cdot$ coefficient of the second term and square the result. 　　　$\frac{1}{2}(-4) = -2$ and $(-2)^2 = \underline{\mathbf{4}}$

　　(-2 is the "half number" and 4 is the "square number")

Write the "square number" 4 in the blank space. 　　　　　　$x^2 - 4x + \underline{\,\mathbf{4}\,}$

The "half number" is -2. Rewrite the trinomial as the square 　　　$= (x - 2)^2$
　of the binomial x $-$ 2.

(c) The coefficient of the second term is 5. To Complete the Square: 　　　$r^2 + 5r + \underline{\,?\,}$

Multiply $\frac{1}{2} \cdot$ coefficient of the second term and square the result. 　　　$\frac{1}{2} \cdot 5 = \frac{5}{2}$ and $\left(\frac{5}{2}\right)^2 = \frac{25}{4}$

　　($\frac{5}{2}$ is the "half number" and $\frac{25}{4}$ is the "square number")

Write the "square number" $\frac{25}{4}$ in the blank space. 　　　　$r^2 + 5r + \underline{\dfrac{25}{4}}$

The "half number" $\frac{5}{2}$ is positive. Rewrite the trinomial as the square 　　$= \left(r + \dfrac{5}{2}\right)^2$

　of the binomial r $+ \dfrac{5}{2}$.

(6.1)

Solving Quadratic Equations – Complete-the-Square Method

EXAMPLE 6   Use the Complete-the-Square Method to solve: $x^2 + 6x + 21 = 0$

Subtract the constant 21 to the right side of the equation. $x^2 + 6x + \underline{\hspace{1cm}} = -21 + \underline{\hspace{1cm}}$
   Add a blank space to both sides of the equation.

Complete the square on the left side of the equation:
   The coefficient of the second term is 6.

$\dfrac{1}{2} \cdot 6 = 3$ and $3^2 = \underline{\mathbf{9}}$

   Add 9 to both sides of the equation. $x^2 + 6x + \underline{\ 9\ } = -21 + \underline{\ 9\ }$

The "half number" 3 is positive. Rewrite the trinomial
   as a (binomial)^2 . Simplify the right side of the equation. $(x + 3)^2 = -12$

Write the square root of both sides of the equation. $\sqrt{(x+3)^2} = \pm\sqrt{-12}$

Use "double signs" for the constant on the right side of the equation.
Simplify both sides of the equation. $x + 3 = \pm\, i\sqrt{4}\sqrt{3}$

$x + 3 = \pm\, 2i\sqrt{3}$

Subtract 3 on both sides of the equation. $\dfrac{-3 \quad\ -3}{}$

Answers: $x = -3 \pm 2i\sqrt{3}$

EXAMPLE 7   Use the Complete-the-Square Method to solve: $w^2 + 7w - 2 = 0$

Add the constant 2 to the right side of the equation. $w^2 + 7w + \underline{\hspace{0.7cm}} = 2 + \underline{\hspace{0.7cm}}$
   Add a blank space to both sides of the equation.

Complete the square on the left side of the equation:
   The coefficient of the second term is 7.

$\dfrac{1}{2}(7) = \dfrac{7}{2}$ and $\left(\dfrac{7}{2}\right)^2 = \dfrac{49}{4}$

   Add $\dfrac{49}{4}$ to both sides of the equation. $w^2 + 7w + \dfrac{49}{4} = 2 + \dfrac{49}{4}$

The "half number" $\dfrac{7}{2}$ is positive. Rewrite the trinomial $\left(w + \dfrac{7}{2}\right)^2 = \left(\dfrac{2}{1}\right)\left(\dfrac{4}{4}\right) + \dfrac{49}{4}$

as a (binomial)^2 . Simplify the right side of the equation. $\left(w + \dfrac{7}{2}\right)^2 = \dfrac{8}{4} + \dfrac{49}{4}$

$\left(w + \dfrac{7}{2}\right)^2 = \dfrac{57}{4}$

Write the square root of both sides of the equation. $\sqrt{\left(w + \dfrac{7}{2}\right)^2} = \pm\dfrac{\sqrt{57}}{\sqrt{4}}$

Use "double signs" for the constant on the right side of the equation.

Simplify both sides of the equation. $w + \frac{7}{2} = \pm\frac{\sqrt{57}}{2}$

Subtract $\frac{7}{2}$ on both sides of the equation. $-\frac{7}{2} \quad -\frac{7}{2}$

Answers: $w = -\frac{7}{2} \pm \frac{\sqrt{57}}{2}$

or $w = \frac{-7 \pm \sqrt{57}}{2}$

Helpful Hint – **Important Rule**: *To complete the square, the leading coefficient must be 1.* If the leading coefficient is not 1, divide the entire equation by the leading coefficient before completing the square.

EXAMPLE 8 Use the Complete-the-Square Method to solve: $3x^2 - 24x - 9 = 0$

The leading coefficient must be 1. Divide the equation by 3. $x^2 - 8x - 3 = 0$

Add the constant 3 to the right side of the equation . $x^2 - 8x + ____ = 3 + ______$
Add a blank space to both sides of the equation.

Complete the square on the left side of the equation:
The coefficient of the second term is – 8.
$\frac{1}{2}(-8) = \mathbf{-4}$ and $(-4)^2 = \underline{\mathbf{16}}$
Add 16 to both sides of the equation. $x^2 - 8x + \underline{\ 16\ } = 3 + \underline{\ 16\ }$

The "half number" – 4 is negative. Rewrite the trinomial as a $(\text{binomial})^2$. Simplify the right side of the equation. $(x-4)^2 = 19$

Write the square root of both sides of the equation. $\sqrt{(x-4)^2} = \pm\sqrt{19}$
Use "double signs" for the constant on the right side of the equation.
Simplify both sides of the equation. $x - 4 = \pm\sqrt{19}$
Add 4 on both sides of the equation. $+4 \quad +4$

Answers: $x = 4 \pm \sqrt{19}$

EXAMPLE 9 Use the Complete-the-Square Method to solve: $2y^2 - 7y + 9 = 0$

The leading coefficient must be 1. Divide the equation by 2. $y^2 - \frac{7}{2}y + \frac{9}{2} = 0$

Subtract the constant $\frac{9}{2}$ to the right side of the equation. $y^2 - \frac{7}{2}y + ___ = -\frac{9}{2} + ____$
Add a blank space to both sides of the equation.

(6.1)

Complete the square on the left side of the equation:

The coefficient of the second term is $-\dfrac{7}{2}$.

$$y^2 - \dfrac{7}{2}y + \underline{\quad} = -\dfrac{9}{2} + \underline{\quad}$$

$$\dfrac{1}{2}\left(-\dfrac{7}{2}\right) = -\dfrac{7}{4} \quad \text{and} \quad \left(-\dfrac{7}{4}\right)^2 = \dfrac{49}{16}$$

Add $\dfrac{49}{16}$ to both sides of the equation.

$$y^2 - \dfrac{7}{2}y + \dfrac{49}{16} = -\dfrac{9}{2} + \dfrac{49}{16}$$

The "half number" $-\dfrac{7}{4}$ is negative. Rewrite the trinomial

$$\left(y - \dfrac{7}{4}\right)^2 = \left(-\dfrac{9}{2}\right)\left(\dfrac{8}{8}\right) + \dfrac{49}{16}$$

as a (binomial)^2. Simplify the right side of the equation.

$$\left(y - \dfrac{7}{4}\right)^2 = -\dfrac{72}{16} + \dfrac{49}{16}$$

$$\left(y - \dfrac{7}{4}\right)^2 = -\dfrac{23}{16}$$

Write the square root of both sides of the equation.

$$\sqrt{\left(y - \dfrac{7}{4}\right)^2} = \pm\sqrt{-\dfrac{23}{16}}$$

Simplify both sides of the equation.

$$\sqrt{\left(y - \dfrac{7}{4}\right)^2} = \pm\dfrac{i\sqrt{23}}{\sqrt{16}}$$

Simplify both sides of the equation.

$$y - \dfrac{7}{4} = \pm\dfrac{i\sqrt{23}}{4}$$

Add $\dfrac{7}{4}$ on both sides of the equation.

$$+\dfrac{7}{4} \qquad +\dfrac{7}{4}$$

Answers:
$$y = \dfrac{7}{4} \pm \dfrac{i\sqrt{23}}{4}$$

$$\text{or} \quad y = \dfrac{7 \pm i\sqrt{23}}{4}$$

6.2 QUADRATIC FORMULA, SUMMARY, APPLICATIONS

QUADRATIC FORMULA

If a quadratic equation is a trinomial of the form $Ax^2 + Bx + C = 0$, then the roots of the equation can be determined by the Quadratic Formula.

> **Quadratic Formula:** If $Ax^2 + Bx + C = 0$ then
> $$x = \frac{-B \pm \sqrt{B^2 - 4AC}}{2A}$$

When using the Quadratic Formula, A is the coefficient of x^2, B is the coefficient of x, and C is the constant term.

�֍ *Helpful Hints* – (1) **Important:** *The Quadratic Formula is for 2nd degree equations only.* Do not use the Quadratic Formula as the first step for solving a 3rd degree or 4th degree equation. Examples for the 3rd degree and 4th degree equations are shown in sections 6.3 and 6.4 of this chapter. (2) To use the Quadratic Formula, the quadratic equation must equal zero.

EXAMPLE 1   Use the Quadratic Formula to solve: $9x^2 - 3x = 1$

The quadratic equation must equal 0.

$$9x^2 - 3x = 1$$

Subtract 1 on both sides of the equation.

$$\underline{-1 \quad -1}$$

$$9x^2 - 3x - 1 = 0$$

Write A, B, and C.

$$A = 9, B = -3, C = -1$$

Set up the Quadratic Formula.

$$x = \frac{-(-3) \pm \sqrt{(-3)^2 - 4(9)(-1)}}{2(9)}$$

Simplify the numerator.

$$= \frac{3 \pm \sqrt{9 + 36}}{18} = \frac{3 \pm \sqrt{45}}{18}$$

Simplify the square root.

$$= \frac{3 \pm \sqrt{9}\sqrt{5}}{18} = \frac{3 \pm 3\sqrt{5}}{18} = \frac{3}{18} \pm \frac{3\sqrt{5}}{18}$$

Then reduce the fraction.

$$x = \frac{1}{6} \pm \frac{\sqrt{5}}{6} \text{ or } \frac{1 \pm \sqrt{5}}{6}$$

EXAMPLE 2   Use the Quadratic Formula to solve: $4x^2 + 3x + 2 = 0$

The quadratic equation already equals zero.

$$4x^2 + 3x + 2 = 0$$

Write A, B, and C.

$$A = 4, B = 3, C = 2$$

Set up the Quadratic Formula.

$$x = \frac{-(3) \pm \sqrt{(3)^2 - 4(4)(2)}}{2(4)}$$

(6.2)

Simplify the numerator. $x = \dfrac{-3 \pm \sqrt{9-32}}{8} = \dfrac{-3 \pm \sqrt{-23}}{8}$

Simplify the square root. $x = \dfrac{-3 \pm i\sqrt{23}}{8}$ or $-\dfrac{3}{8} \pm \dfrac{i\sqrt{23}}{8}$

Helpful Hints – SUMMARY: SOLVING SECOND-DEGREE EQUATIONS

To solve a **quadratic equation**, use the shortest, most efficient method:

(1) Use the Factoring Method *whenever it is possible.*

(2) Use the Square Root Method for ***a binomial** quadratic equation which cannot be factored*.

(3) Use the Quadratic Formula for ***a trinomial** quadratic equation which cannot be factored*. The Complete-the-Square Method may also be used for a trinomial quadratic equation which cannot be factored.

The following practice problems in Example 3 illustrate how to apply the most efficient method for solving a quadratic equation.

EXAMPLE 3 Name the shortest, most efficient method and solve each equation.

(a) $w^2 + 4w = 0$ (b) $x^2 + 4 = 0$ (c) $y^2 - 4y + 3 = 0$ (d) $z^2 - 4z + 1 = 0$

Answers:

(a) The most efficient method is to solve this quadratic equation by Factoring. $w^2 + 4w = 0$

Factor the polynomial by using the Greatest Common Factor method. $w(w + 4) = 0$

Solve for *w*. $w = 0$ or $w = -4$

(b) The binomial $x^2 + 4$ cannot be factored. Use the Square Root Method to solve the quadratic equation. $x^2 + 4 = 0$

$-4 \quad -4$

Subtract 4 on both sides of the equation. $x^2 = -4$

Write the square root of both sides of the equation. $\sqrt{x^2} = \pm\sqrt{-4}$

Simplify both sides of the equation. $x = \pm 2i$

(c) The most efficient method is to solve this quadratic equation by Factoring. $y^2 - 4y + 3 = 0$

Factor the polynomial by using the Trinomial method. $(y - 3)(y - 1) = 0$

Solve for *y*. $y = 3$ or $y = 1$

(6.2)

(d) The trinomial $z^2 + 4z + 1$ cannot be factored. Use the Quadratic
    Formula to solve the quadratic equation. $z^2 - 4z + 1 = 0$

Write A, B, and C. $A = 1, B = -4, C = 1$

Set up the Quadratic Formula. $z = \dfrac{-(-4) \pm \sqrt{(-4)^2 - 4(1)(1)}}{2(1)}$

Simplify the formula. $= \dfrac{4 \pm \sqrt{16 - 4}}{2} = \dfrac{4 \pm \sqrt{12}}{2}$

$$= \dfrac{4 \pm \sqrt{4}\sqrt{3}}{2} = \dfrac{4 \pm 2\sqrt{3}}{2} = \dfrac{4}{2} \pm \dfrac{2\sqrt{3}}{2}$$

$$z = 2 \pm \sqrt{3}$$

APPLICATIONS

For each word problem, write the equation and find the original **exact roots** of the equation (the roots determined by algebraic procedures). Then use a calculator and round each answer to a decimal.

EXAMPLE 4   An object is fired from ground level. After **t** seconds, its height in feet above the ground is given by $H = 10t - t^2$. At what times is the object 10 feet above the ground? (Give the answers to the nearest tenth.)

The height H = 10 feet. Substitute 10 in place of H in the formula. $10 = 10t - t^2$

Since the leading term $-t^2$ is negative, cancel out
   the right side of the equation. $-10t + t^2$ $-10t + t^2$

The trinomial $t^2 - 10t + 10$ cannot be factored.

Use the Quadratic Formula Method to solve the quadratic equation. $t^2 - 10t + 10 = 0$

  Write A, B, and C. $A = 1, B = -10, C = 10$

  Set up the Quadratic Formula. $t = \dfrac{-(-10) \pm \sqrt{(-10)^2 - 4(1)(10)}}{2(1)}$

  Simplify the numerator. $= \dfrac{10 \pm \sqrt{100 - 40}}{2} = \dfrac{10 \pm \sqrt{60}}{2}$

  Simplify the square root. $= \dfrac{10 \pm \sqrt{4}\sqrt{15}}{2} = \dfrac{10 \pm 2\sqrt{15}}{2}$

  Reduce the fraction. $t = \dfrac{10}{2} \pm \dfrac{2\sqrt{15}}{2} = 5 \pm \sqrt{15}$

                            Exact roots: $t = 5 - \sqrt{15}$ or $t = 5 + \sqrt{15}$

Use a calculator to evaluate each of the roots, rounded to the nearest tenth.
Decimal results: $t = 5 - \sqrt{15} \approx 1.1$ seconds; $t = 5 + \sqrt{15} \approx 8.9$ seconds

Answers: When the object is moving up in the air, it will be 10 feet above the ground at t = 1.1 seconds. When the object is coming down, it will be 10 feet above the ground at time t = 8.9 seconds.

(6.2)

EXAMPLE 5   The legs of the right triangle at the right are of equal length. The hypotenuse is 6 inches long. Find the length of each leg of the triangle. (Give the answers to the nearest hundredth.)

Table   (Set-up)

Leg A   x

Leg B   x   (Same length as leg A)

Hypotenuse C   6 inches

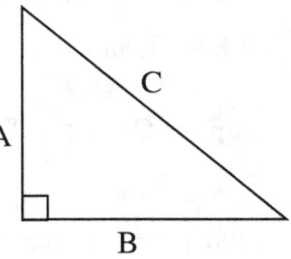

Formula for the sides of the right triangle: $A^2 + B^2 = C^2$

Equation: $x^2 + x^2 = 6^2$

Simplify the equation. Solve by using the Square Root Method. $2x^2 = 36$

Solve for x^2 by dividing both sides of the equation by 2. $x^2 = 18$

Write the square root of both sides of the equation. $\sqrt{x^2} = \pm\sqrt{18}$

Simplify both sides of the equation. $x = \pm\sqrt{9}\sqrt{2} = \pm 3\sqrt{2}$

i.e. $x = -3\sqrt{2}$ or $x = 3\sqrt{2}$

The sides of a geometric figure must be positive.
Delete the negative answer.   Exact root: $x = 3\sqrt{2}$

Use a calculator to evaluate the exact root, rounded to the nearest hundredth: $x = 3\sqrt{2} \approx 4.24$ in.

Answer: Each leg of the right triangle is 4.24 inches.

EXAMPLE 6   Working together, it takes Tim and Ed 3 hours to replace the wall-to-wall carpet in a large den. If Tim works alone, it will take him one hour longer to do the job than it would take Ed. How long will it take for Tim to replace the carpet?

We will use the formula given in Section 4.5 for two people working together.

Workers   Rates

Tim   x + 1

Ed   x

Together   3 hrs

Formula for Work Problems: $\dfrac{\text{rate together}}{\text{1st rate}} + \dfrac{\text{rate together}}{\text{2nd rate}} = 1$ job

Use the formula to write the equation. $\dfrac{3}{x+1} + \dfrac{3}{x} = 1$

In this equation, $x + 1 \neq 0$ and $x \neq 0$. Restricted values: $x \neq -1, x \neq 0$

Multiply the LCD (x + 1)(x) times each term of the equation.

$$(x+1)(x)\frac{3}{x+1} + (x+1)(x)\frac{3}{x} = 1(x+1)(x)$$

Cancel each denominator. $$\cancel{(x+1)}(x)\frac{3}{\cancel{x+1}} + (x+1)\cancel{(x)}\frac{3}{\cancel{x}} = 1(x+1)(x)$$

Multiply the result times the numerator. $3x + 3x + 3 = x^2 + x$

Combine the similar terms. $6x + 3 = x^2 + x$

The leading term x^2 is positive. $-6x - 3 \quad -6x - 3$

Cancel the left side of the equation. $0 = x^2 - 5x - 3$

The trinomial cannot be factored. Use the Quadratic Formula Method. $A = 1, B = -5, C = -3$

Set up the Quadratic Formula. $$x = \frac{-(-5) \pm \sqrt{(-5)^2 - 4(1)(-3)}}{2(1)}$$

Simplify the formula. $$= \frac{5 \pm \sqrt{25+12}}{2} = \frac{5 \pm \sqrt{37}}{2}$$

Exact roots: $x = \frac{5-\sqrt{37}}{2}$ or $x = \frac{5+\sqrt{37}}{2}$

Use a calculator to evaluate each of the roots, rounded to the nearest tenth.

Decimal results: $x = \frac{5-\sqrt{37}}{2} \approx -0.5$ Time is positive; delete the negative result.

$x = \frac{5+\sqrt{37}}{2} \approx 5.5$ hours (For Ed) $x + 1 = 5.5 + 1 \approx 6.5$ hours (For Tim)

Answers: Working alone, it will take Ed 5.5 hours to replace the carpet; it will take Tim 6.5 hours to replace the carpet.

6.3 SOLVING POLYNOMIAL EQUATIONS (3RD AND 4TH DEGREE)

In Section 3.4 we learned that the degree of a polynomial equation indicates the exact number of roots of the equation. Thus a cubic equation is a third-degree polynomial equation and it has exactly 3 roots. A fourth-degree polynomial equation has exactly 4 roots or solutions.

Helpful Hint – **Important:** *To solve a polynomial equation (which has an exponent), set the equation equal to zero.* Then determine whether the polynomial can be factored and use the most efficient method(s) to solve the equation.

SOLVING A POLYNOMIAL EQUATION WITH LINEAR FACTORS

EXAMPLE 1 Solve the polynomial equation: $50x^3 = 8x$

The polynomial equation must equal zero.

$$50x^3 = 8x$$
$$-8x \quad -8x$$
$$50x^3 - 8x = 0$$

(6.3)

Factor the polynomial using the Greatest Common Factor method. $2x(25x^2 - 4) = 0$

Factor the polynomial using the SQ – SQ method. $2x(5x - 2)(5x + 2) = 0$

Solve for x. $x = 0, \quad x = \dfrac{2}{5}, \quad x = -\dfrac{2}{5}$

EXAMPLE 2   Solve the polynomial equation: $x^4 - 17x^2 + 16 = 0$

The polynomial equation already equals zero. $x^4 - 17x^2 + 16 = 0$

Factor the polynomial using the Trinomial method   . $(x^2 - 1)(x^2 - 16) = 0$

Factor each binomial using the SQ – SQ method. $(x - 1)(x + 1)(x - 4)(x + 4) = 0$

Solve for x. $x = 1, \quad x = -1, \quad x = 4, \quad x = -4$

SOLVING A POLYNOMIAL EQUATION WITH 2ND-DEGREE FACTORS

> **To solve a 3rd or 4th degree equation:**
>
> - The equation must equal zero.
>
> - Factor the polynomial (using the Greatest Common Factor, Square – Square, Trinomial, or the Grouping methods).
>
> - Set each factor which contains a variable equal zero and solve for x.
>
> - For the 2nd -degree equation(s), use the most appropriate (shortest) method for solving the equation (Factoring, Square Root method, or the Quadratic Formula method).

EXAMPLE 3   Solve the polynomial equation: $x^3 - 3x^2 - x = 0$

The polynomial equation already equals zero. $x^3 - 3x^2 - x = 0$

Factor the polynomial using the Greatest Common Factor method. $x(x^2 - 3x - 1) = 0$

Set each factor which contains a variable equal zero and solve for x. $x = 0$

$x^2 - 3x - 1 = 0$

The trinomial cannot be factored. Use the Quadratic Formula Method. $A = 1, B = -3, C = -1$

Write A, B, and C. Set up the formula. $x = \dfrac{-(-3) \pm \sqrt{(-3)^2 - 4(1)(-1)}}{2(1)}$

Simplify the numerator. $x = \dfrac{3 \pm \sqrt{9 + 4}}{2} = \dfrac{3 \pm \sqrt{13}}{2}$

Answers: $x = 0, \quad x = \dfrac{3 \pm \sqrt{13}}{2}$

EXAMPLE 4   Solve the polynomial equation: $x^3 - 4x^2 + 25x - 100 = 0$

The polynomial equation already equals zero.	$x^3 - 4x^2 + 25x - 100 = 0$
Factor the polynomial using the Grouping method (two answer lines).	$x^2(x - 4) + 25(x - 4) = 0$
	$(x - 4)(x^2 + 25) = 0$
Set each factor which contains a variable equal zero.	$x - 4 = 0 \qquad x^2 + 25 = 0$
Solve for x.	$x = 4$

The binomial $x^2 + 25$ cannot be factored. Use the Square Root Method. Subtract 25 on both sides of the equation.

$$x^2 = -25$$

Write the square root of both sides of the equation.

$$\sqrt{x^2} = \pm\sqrt{-25}$$

Simplify both sides of the equation.

$$x = \pm 5i$$

Answers: $x = 4$, $x = \pm 5i$

EXAMPLE 5   Solve the polynomial equation: $2x^4 - 3x^2 - 2 = 0$

The polynomial equation already equals zero.	$2x^4 - 3x^2 - 2 = 0$
Factor the polynomial using the Trinomial method.	$(2x^2 + 1)(x^2 - 2) = 0$
Set each factor which contains a variable equal zero. The binomials cannot be factored.	$2x^2 + 1 = 0 \quad\bigg|\quad x^2 - 2 = 0$ $2x^2 = -1$

Solve each equation using the Square Root Method.

$$x^2 = -\frac{1}{2} \qquad\bigg|\qquad x^2 = 2$$

Write the square root of both sides of the equation.

$$\sqrt{x^2} = \pm\sqrt{-\frac{1}{2}} \qquad\bigg|\qquad \sqrt{x^2} = \pm\sqrt{2}$$

Simplify both sides of the equation.

$$x = \pm i\frac{\sqrt{1}}{\sqrt{2}} \cdot \frac{\sqrt{2}}{\sqrt{2}} \qquad\bigg|\qquad x = \pm\sqrt{2}$$

$$x = \pm i\frac{\sqrt{2}}{2}$$

Answers: $x = \pm i\dfrac{\sqrt{2}}{2}$, $x = \pm\sqrt{2}$

(6.3)

EXAMPLE 6   Solve the polynomial equation: $80x^4 + 270x = 0$

The polynomial equation already equals zero. $80x^4 + 270x = 0$

Factor the polynomial using the Greatest Common Factor method. $10x(8x^3 + 27) = 0$

Factor the polynomial using the Sum of Cubes method.

"$8x^3 + 27$" will factor as a (binomial)(trinomial):

To get the binomial factor, identify the base of each cubic term. $10x(2x + 3)(\quad ? \quad) = 0$

Since $8x^3 = (2x)^3$, the base for $8x^3$ is 2x. Since $27 = (3)^3$, the base for 27 is 3.

Use the *binomial* to set up the trinomial factor. $10x(2x + 3)[(2x)^2 - (2x)(3) + (3)^2] = 0$

"SQ" - - Square the first term of the binomial. $\qquad$ SQ $\quad$ Mult $\quad$ SQ

"Mult & Chg Sgn" - - Multiply both terms and change the sign. $\qquad\qquad$ & Chg Sgn

"SQ" - - Square the second term of the binomial.

Simplify the trinomial factor. $10x(2x + 3)(4x^2 - 6x + 9) = 0$

Set each factor which contains a variable equal zero and solve for x. $10x = 0 \ \rightarrow \ x = 0$

$$2x + 3 = 0 \ \rightarrow \ x = -\frac{3}{2}$$

The trinomial cannot be factored. Use the Quadratic Formula Method. $4x^2 - 6x + 9 = 0$

Write A, B, and C. Set up the formula. $A = 4, B = -6, C = 9$

$$x = \frac{-(-6) \pm \sqrt{(-6)^2 - 4(4)(9)}}{2(4)}$$

Simplify the numerator. $x = \dfrac{6 \pm \sqrt{36 - 144}}{8}$

$$x = \frac{6 \pm \sqrt{-108}}{8}$$

To reduce the square root: $108 = 9 \cdot 12 = 9 \cdot 4 \cdot 3 = 36 \cdot 3$ $x = \dfrac{6 \pm i\sqrt{36}\,\sqrt{3}}{8}$

Simplify the numerator. $x = \dfrac{6 \pm 6i\sqrt{3}}{8}$

Reduce the fraction. (Divide each term by 2.) $x = \dfrac{3 \pm 3i\sqrt{3}}{4}$

Answers: $x = 0$, $x = -\dfrac{3}{2}$, $x = \dfrac{3 \pm 3i\sqrt{3}}{4}$

6.4   EQUATIONS IN QUADRATIC FORM   (SUBSTITUTION METHOD)

The examples in the box at the right are **equations in quadratic form.** Each equation is written in a "trinomial" format: $a(\ldots)^2 + b(\ldots) + c = 0$

Also observe:

x^4 is the square of x^2 because $(x^2)^2 = x^4$.

$v^{2/3}$ is the square of $v^{1/3}$ because $[v^{1/3}]^2 = v^{2/3}$.

$(y-1)^{-2}$ is the square of $(y-1)^{-1}$ because $\left[(y-1)^{-1}\right]^2 = (y-1)^{-2}$.

> *Examples of*
> *Equations in Quadratic Form*
>
> 1. $4x^4 - 19x^2 - 5 = 0$
> 2. $v^{2/3} - 3v^{1/3} + 2 = 0$
> 3. $(y-1)^{-2} - 2(y-1)^{-1} - 8 = 0$

Use the **Substitution Method** to solve equations in quadratic form.

> Steps:
> * The equation must equal zero.
> * Choose a variable that is NOT used in the equation (such as "z"). Let z represent the variable expression of the second term. Let z^2 represent the variable expression of the first term.
>
> $$a(\ldots) + b(\ldots) + c = 0$$
> $$\uparrow_{z^2} \qquad \uparrow_{z}$$
>
> * Using substitution, rewrite the equation as a quadratic equation in terms of z.
> * Solve the new equation for z.
> * Use substitution to replace each z with the original variable expression.
> * Now solve each equation for the original variable.

EXAMPLE 1   Solve by using the Substitution Method: $4x^4 - 19x^2 - 5 = 0$

The equation already equals zero. $4x^4 - 19x^2 - 5 = 0$

Choose a variable that is NOT used in the equation (such as "z"). $z = x^2$ and $z^2 = x^4$

  Let $z = x^2$ and $z^2 = x^4$. Rewrite the equation in terms of z. $4z^2 - 19z - 5 = 0$

Factor the trinomial. $(4z + 1)(z - 5) = 0$

Solve for z. $z = -\dfrac{1}{4}$ $\bigg|$ $z = 5$

Substitute the original expression x^2 in place of z. $x^2 = -\dfrac{1}{4}$ $\bigg|$ $x^2 = 5$

Solve each equation for x using the Square Root Method. $\sqrt{x^2} = \pm\sqrt{-\dfrac{1}{4}}$ $\bigg|$ $\sqrt{x^2} = \pm\sqrt{5}$

$$x = \pm i \dfrac{\sqrt{1}}{\sqrt{4}} \quad \bigg| \quad x = \pm\sqrt{5}$$

Answers: $x = \pm\dfrac{1}{2}i$, $x = \pm\sqrt{5}$

(6.4)

✳ **_Helpful Hint_** – Remember: When solving an equation in quadratic form using the Substitution Method, *you must choose a variable that is **not** used in the equation.*

EXAMPLE 2   Solve by using the Substitution Method: $v^{2/3} = 3v^{1/3} - 2$

The equation must equal zero.
$$v^{2/3} - 3v^{1/3} + 2 = 0$$

Choose a variable that is NOT used in the equation (such as "z").
$$z = v^{1/3} \text{ and } z^2 = v^{2/3}$$

Let $z = v^{1/3}$ and $z^2 = v^{2/3}$. Rewrite the equation in terms of z.
$$z^2 - 3z + 2 = 0$$

Factor the trinomial.
Solve for z.
$$(z - 1)(z - 2) = 0$$
$$z = 1 \quad | \quad z = 2$$

Substitute the original expression $v^{1/3}$ in place of z.
$$v^{1/3} = 1 \quad | \quad v^{1/3} = 2$$

Write the left side of the equation as a cube root.
$$\sqrt[3]{v} = 1 \quad | \quad \sqrt[3]{v} = 2$$

To solve for v, cube both sides of the equation.
$$\left(\sqrt[3]{v}\right)^3 = (1)^3 \quad \left| \quad \left(\sqrt[3]{v}\right)^3 = (2)^3 \right.$$

Answers: $v = 1$, $v = 8$

EXAMPLE 3   Solve by using the Substitution Method: $(y-1)^{-2} - 2(y-1)^{-1} = 8$

The equation must equal zero.
$$(y-1)^{-2} - 2(y-1)^{-1} - 8 = 0$$

Choose a variable that is NOT used in the equation (such as "z").
$$z = (y-1)^{-1} \text{ and } z^2 = (y-1)^{-2}$$

Let $z = (y-1)^{-1}$ and $z^2 = (y-1)^{-2}$. Rewrite the equation in terms of z.
$$z^2 - 2z - 8 = 0$$

Factor the trinomial.
Solve for z.
$$(z + 2)(z - 4) = 0$$
$$z = -2 \quad | \quad z = 4$$

Substitute the original expression $(y-1)^{-1}$ in place of z.
$$(y-1)^{-1} = -2 \quad | \quad (y-1)^{-1} = 4$$

Write both sides of the equation in fraction form.
$$\frac{1}{y-1} = \frac{-2}{1} \quad \left| \quad \frac{1}{y-1} = \frac{4}{1} \right.$$

State the restricted value of the denominator: $y \neq 1$.

The LCD is $y - 1$.

Multiply $(y - 1)$ times the equation.
$$(y-1)\frac{1}{y-1} = \frac{-2}{1}(y-1) \quad \left| \quad (y-1)\frac{1}{y-1} = \frac{4}{1}(y-1) \right.$$

Rewrite the equation and solve for y.
$$1 = -2y + 2 \quad \left| \quad 1 = 4y - 4 \right.$$
$$-1 = -2y \quad \left| \quad 5 = 4y \right.$$

Answers: $y = \dfrac{1}{2}$, $y = \dfrac{5}{4}$

6.5   GRAPHING QUADRATIC FUNCTIONS

The standard format for a **quadratic function** is $f(x) = ax^2 + bx + c$ or $y = ax^2 + bx + c$. The graph of a quadratic function is a vertical **parabola**, as shown below. (Also refer to Section 2.1, Example 3.)

If the leading coefficient $a > 0$, the parabola opens up and the **vertex** is the minimum (lowest) point of the graph. If $a < 0$, the parabola opens down and the vertex is the maximum (highest) point of the graph.

The **axis of symmetry** is a dotted vertical line through the vertex. A parabola is symmetric or balanced on both sides of the axis of symmetry so that the right half of the parabola is a mirror image of the left half.

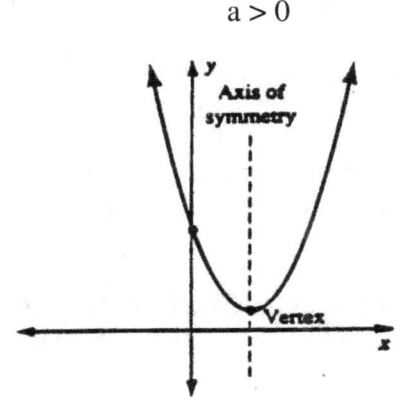

$a > 0$

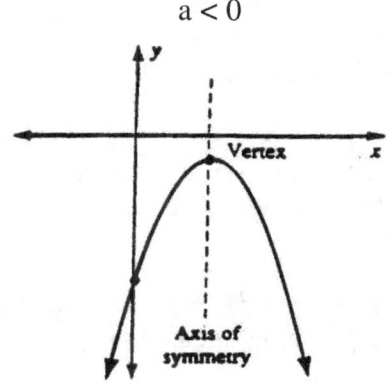

$a < 0$

MONOMIAL QUADRATIC FUNCTIONS

> **To Graph The Function $f(x) = ax^2$:**
> - The *vertex* is $(0, 0)$.
> - Make a *5-point table* of ordered pairs using the integers of $[-2, 2]$ in the **x** column.
> - Neatly draw the *graph*, rounded at the vertex.

EXAMPLE 1   Graph $f(x) = -x^2$ by making a 5-point table of ordered pairs and using the integers of $[-2, 2]$ in the x column. Also, identify the vertex.

Answer: The leading coefficient of $f(x) = -x^2$ is $a = -1$, which indicates that the graph will open downward.

x	$f(x) = -x^2$	(Workspace)
-2	-4	$-(-2)^2 = -(4)$
-1	-1	$-(-1)^2 = -(1)$
0	0	$-(0)^2 = 0$
1	-1	$-(1)^2 = -(1)$
2	-4	$-(2)^2 = -(4)$

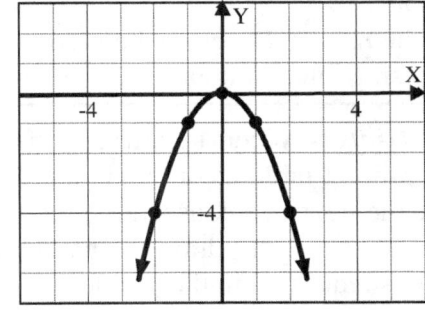

The vertex is maximum point $(0, 0)$.

127

(6.5)

✳ *__Helpful Hint__* – When graphing a parabola, the *vertex must be rounded*, not a corner point. In general, the graph of a monomial or polynomial function has *no corner points*!

QUADRATIC FUNCTIONS IN "VERTEX FORMAT"

The **vertex format** for a quadratic function is $f(x) = a(x - h)^2 + k$. Observe that the function involves a binomial squared. The vertex is indicated by the function itself.

- The leading coefficient *a* indicates whether the parabola opens up or opens down.
- The *vertex = (h, k)*: *h* uses the opposite sign of the inside constant and *k* is the given outside constant.

EXAMPLE 2   For each quadratic function, find the vertex without drawing the graph. Determine whether the vertex is the minimum or maximum point of the graph.

(a) $y = 2(x - 1)^2 + 5$ (b) $f(x) = -2(x + 3)^2 + 4$ (c) $f(x) = (x + 2)^2 - 5$

Answers:

(a) For $y = 2(x - 1)^2 + 5$, the square indicates that the graph is a parabola. The leading coefficient is $a = 2$; the parabola opens up. Also, $h = 1$ and $k = 5$; the vertex $(1, 5)$ is the minimum point of the parabola.

(b) For $f(x) = -2(x + 3)^2 + 4$, the square indicates that the graph is a parabola. The leading coefficient is $a = -2$; the parabola opens down. Also, $h = -3$ and $k = 4$; the vertex $(-3, 4)$ is the maximum point of the parabola.

(c) For $f(x) = (x + 2)^2 - 5$, the square indicates that the graph is a parabola. The leading coefficient is $a = 1$ (in front of the binomial squared); the parabola opens up. Also, $h = -2$ and $k = -5$; the vertex $(-2, -5)$ is the minimum point of the parabola.

Steps for Graphing a Quadratic Function $f(x) = a(x - h)^2 + k$
- State the vertex (h, k).
- Make a *5-point table* of ordered pairs. Draw the graph. (See the *Helpful Hints* below.)
- The *axis of symmetry* is the vertical line $x = h$ which passes through the vertex.
- The *y – intercept = (0, y)*: Let $x = 0$ and find the value of y.
- The *x – intercept(s) = (x, 0)*: Let $y = 0$ and find the value of x. (If the answers for x are imaginary, there are no x – intercepts!)
- State the *domain*: Describe the x-coordinates of the graph from left to right using interval notation.
- State the *range*: Describe the y-coordinates of the graph from the lowest point to the highest point using interval notation.

✳ *__Helpful Hints__* – When identifying the parts of a quadratic function: (1) Determine the vertex. (2) Make a 5-point table of ordered pairs by placing the vertex in the middle of the table. For the x-column, use the next two consecutive integers to the left and to the right of the vertex. Determine the y-coordinates. (3) Neatly draw the parabola. (4) Use the graph and the above answers to determine the other items requested in the problem.

The following examples will illustrate the steps listed above.

EXAMPLE 3   For the function $f(x) = (x - 2)^2 - 1$, determine: (a) the vertex, (b) the maximum or minimum point, (c) a 5-point table and the graph, (d) the axis of symmetry, (e) the intercepts, (f) the domain, (g) the range.

Answers:

(a) For $f(x) = (x - 2)^2 - 1$, the square indicates that the graph is a parabola.   Also, h = 2 and k = – 1. The vertex is (2, – 1).

(b) The leading coefficient is a = 1 (in front of the binomial squared). The parabola opens up. The vertex (2, – 1) is the minimum point of the parabola.

(c) Set up the 5-point table by placing the vertex in the middle of the table. For the x-column, use the next two consecutive integers to the left and to the right of the vertex. Find the y-coordinates.

Table   Graph

x	$f(x) = (x - 2)^2 - 1$	(Workspace)
0	3	$(0 - 2)^2 - 1 = (- 2)^2 - 1 = 4 - 1$
1	0	$(1 - 2)^2 - 1 = (- 1)^2 - 1 = 1 - 1$
2	– 1	Vertex
3	0	$(3 - 2)^2 - 1 = (1)^2 - 1 = 1 - 1$
4	3	$(4 - 2)^2 - 1 = (2)^2 - 1 = 4 - 1$

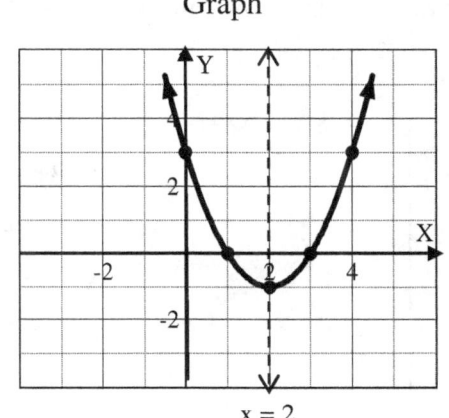

x = 2

(d) From answer (a) above, h = 2. The axis of symmetry is the vertical line x = 2, which passes through the vertex.

(e) Referring to the table or the graph, the x-intercepts are (1, 0) and (3, 0). The y-intercept is (0, 3).

(f) The left end of the graph has an arrow which indicates – ∞. The right end of the graph has an arrow which indicates ∞. For $f(x) = (x - 2)^2 - 1$, the domain = {all real numbers} = (– ∞, ∞).

(g) The lowest y-coordinate of the graph is – 1. The highest point of the graph has arrows which indicate ∞. For $f(x) = (x - 2)^2 - 1$, the range = [– 1, ∞).

QUADRATIC FUNCTIONS IN STANDARD FORM

The **standard form** for a quadratic function is $f(x) = ax^2 + bx + c$. The function is given in the traditional polynomial format and the vertex is <u>not</u> given. To determine the vertex, use the following formulas:

Given the quadratic function $y = f(x) = ax^2 + bx + c$:
The **vertex = (h, k)** where $h = \dfrac{-b}{2a}$ and $k = f(h)$.
(To find k, substitute h in place of each x in the function and simplify.)

(6.5)

EXAMPLE 4 For the function $y = x^2 + 2x - 3$, determine:
(a) the vertex, (b) a 5-point table and the graph, (c) the axis of symmetry, (d) the intercepts.

Answers: Remember that $y = x^2 + 2x - 3$ can be written using function notation as $f(x) = x^2 + 2x - 3$.

(a) For $y = x^2 + 2x - 3$, the square indicates that the graph is a parabola. The leading coefficient a = 1 indicates that the parabola opens up. The vertex is not given in the function. Also, a = 1, b = 2, and

c = − 3. The vertex = (h, k) where $h = \dfrac{-b}{2a} = \dfrac{-2}{2(1)} = -1$. To find k, substitute h = − 1 in place of each

x in the function to get $k = f(-1) = (-1)^2 + 2(-1) - 3 = 1 - 2 - 3 = -4$. The vertex is (− 1, − 4), the minimum point of the parabola.

(b) Set up the 5-point table by placing the vertex in the middle of the table. For the x-column, use the next two consecutive integers to the left and to the right of the vertex. Find the y-coordinates.

Table   Graph

x	$y = x^2 + 2x - 3$	(Workspace)
− 3	0	$(-3)^2 + 2(-3) - 3 = 9 - 6 - 3$
− 2	− 3	$(-2)^2 + 2(-2) - 3 = 4 - 4 - 3$
− 1	− 4	Vertex
0	− 3	$(0)^2 + 2(0) - 3 = 0 + 0 - 3$
1	0	$(1)^2 + 2(1) - 3 = 1 + 2 - 3$

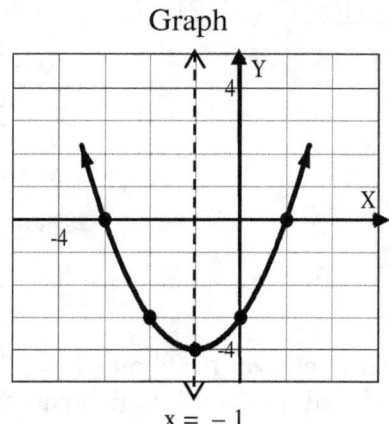

x = − 1

(c) From answer (a) above, h = − 1. The axis of symmetry is the vertical line x = − 1, which passes through the vertex.

(d) Referring to the graph, the x-intercepts are (− 3, 0) and (1, 0). The y-intercept is (0, − 3).

EXAMPLE 5 For the function $f(x) = -x^2 + 6$, determine: (a) the vertex, (b) a 5-point table and the graph, (c) the intercepts, (d) the domain, (e) the range.

Answers:

(a) For $f(x) = -x^2 + 6$, the square indicates that the graph is a parabola. Also, a = − 1, b = 0, and c = 6. The leading coefficient a = − 1 indicates that the parabola opens down. The vertex is not given in the

function. The vertex = (h, k) where $h = \dfrac{-b}{2a} = \dfrac{-0}{2(-1)} = 0$. To find k, substitute h = 0 in place of each

x in the function to get $k = f(0) = -(0)^2 + 6 = 0 + 6 = 6$. The vertex is (0, 6), the maximum point of the parabola.

(b) Set up the 5-point table by placing the vertex in the middle of the table. For the x-column, use the next two consecutive integers to the left and to the right of the vertex. Find the y-coordinates.

(6.5)

Table | | | Graph

x	$f(x) = -x^2 + 6$	(Workspace)
-2	2	$-(-2)^2 + 6 = -4 + 6$
-1	5	$-(-1)^2 + 6 = -1 + 6$
0	6	Vertex
1	5	$-(1)^2 + 6 = -1 + 6$
2	2	$-(2)^2 + 6 = -4 + 6$

From answer (a), h = 0. The axis of symmetry is the vertical line x = 0, which is the y-axis.

(c) Referring to the graph, the x-intercepts are between the vertical grid lines. To find the x-intercepts, let y = 0 and solve for x:

Write the function.	$f(x) = -x^2 + 6$
Change the f(x) symbol to y.	$y = -x^2 + 6$
Substitute 0 in place of y.	$0 = -x^2 + 6$
Add x^2 to both sides of the equation.	$x^2 = 6$
Write the square root of both sides.	$\sqrt{x^2} = \pm\sqrt{6}$
Simplify both sides.	$x = \pm\sqrt{6}$
Use a calculator to evaluate the answers for x.	$x \approx \pm 2.45$

The x-intercepts are (– 2.45, 0) and (2.45, 0).

Referring to the graph, the y-intercept is (0, 6).

(d) The left end of the graph has an arrow which indicates – ∞. The right end of the graph has an arrow which indicates ∞. For $f(x) = -x^2 + 6$, the domain = {all real numbers} = (– ∞, ∞).

(e) The lowest point of the graph has arrows which indicate – ∞. The highest y-coordinate of the graph is 6.   For $f(x) = -x^2 + 6$, the range = (– ∞, 6].

EXAMPLE 6   For the function $f(x) = -2x^2 + 6x - 5$, determine: (a) the vertex, (b) a 5-point table and the graph, (c) axis of symmetry, (d) the intercepts, (e) the domain, (f) the range.

Answers:

(a) For $f(x) = -2x^2 + 6x - 5$, the square indicates that the graph is a parabola. Also, a = – 2, b = 6, and c = – 5.   The leading coefficient a = – 2 indicates that the parabola opens down. The vertex is not given in the function. The vertex = (h, k) where $h = \dfrac{-b}{2a} = \dfrac{-6}{2(-2)} = \dfrac{-6}{-4} = 1.5$. To find k, substitute h = 1.5 in place of each x in the function to get k = f(1.5) = $-2(1.5)^2 + 6(1.5) - 5 = -0.5$. The vertex is (1.5, – 0.5), the maximum point of the parabola.

(b) Set up the 5-point table by placing the vertex in the middle of the table. For the x-column, use the next two consecutive integers to the left and to the right of the vertex. Find the y-coordinates.

131

(6.5)

Table

Graph

x	$f(x) = -2x^2 + 6x - 5$	(Workspace)
0	−5	$-2(0)^2 + 6(0) - 5 = 0 + 0 - 5$
1	−1	$-2(1)^2 + 6(1) - 5 = -2 + 6 - 5$
1.5	−0.5	Vertex
2	−1	$-2(2)^2 + 6(2) - 5 = -8 + 12 - 5$
3	−5	$-2(3)^2 + 6(3) - 5 = -18 + 18 - 5$

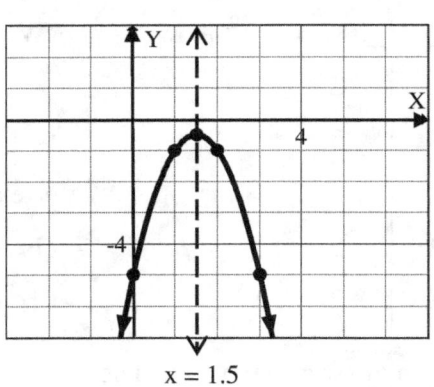

x = 1.5

(c) From answer (a), h = 1.5. The axis of symmetry is the vertical line x = 1.5.

(d) Referring to the graph, there are no x-intercepts. The y-intercept is (0, − 5).

(e) The left end of the graph has an arrow which indicates − ∞. The right end of the graph has an arrow which indicates ∞. For $f(x) = -2x^2 + 6x - 5$, the domain = {all real numbers} = (− ∞, ∞).

(f) The lowest point of the graph has arrows which indicate − ∞. The highest y-coordinate of the graph is − 0.5.   For $f(x) = -2x^2 + 6x - 5$, the range = (− ∞, − 0.5].

APPLICATIONS

For word problems which involve a quadratic function, the words "maximum" or "minimum" refer to the vertex of the parabola. Find the vertex of the function and use the coordinates of the vertex to determine the answers to the word problem.

EXAMPLE 7   Finding the Maximum Value: A farmer estimates the **profit** P in dollars from grazing x cattle on a pasture is given by $P(x) = -3.2x^2 + 352x - 3600$. Find the following answers without drawing the graph.
  (a) Find the profit if the farmer has 25 cattle grazing on the pasture.
  (b) How many cattle should be grazed on the pasture if the farmer wants a profit of $5,360?
  (c) How many cattle should be grazed on this pasture to get the maximum profit? What is the maximum profit?

Answers:

(a) Let x = 25 cattle. $P(25) = -3.2(25)^2 + 352(25) - 3600 = -3.2(625) + 8800 - 3600 = 3200$.
The profit is $3,200.

(b) Let P = $5,360. Substitute $5,360 in place of P(x) and solve the equation for x.

The equation has an exponent.	$5360 = -3.2x^2 + 352x - 3600$
To set the equation equal to zero, cancel out the right side of the equation.	$3.2x^2 - 352x + 3600 + 5360 = 0$ $3.2x^2 - 352x + 8960 = 0$
Multiply each term times 10 to eliminate the decimal.	$32x^2 - 3520x + 89{,}600 = 0$

Factor out the greatest common factor, 32.	$32(x^2 - 110x + 2800) = 0$
Factor the trinomial.	$32(x - 40)(x - 70) = 0$
Solve for x.	$x = 40$ or $x = 70$

Observe that the *same profit* of $5,360 will be gained from grazing 40 cattle or 70 cattle. However, it takes less effort by the farmer to manage 40 cattle than 70 cattle. Therefore, 40 cattle should graze on the pasture if the farmer wants a profit of $5,360.

(c) $P(x) = -3.2x^2 + 352x - 3600$ is a second-degree function whose graph is a parabola. The leading coefficient is a = -3.2 which indicates that the parabola opens downward. The maximum point of the function is the vertex. To determine the maximum profit, we will find the vertex.

For this function, a = -3.2, b = 352, and c = -3600.

The vertex = (h, k) where $h = \dfrac{-b}{2a} = \dfrac{-352}{2(-3.2)} = \dfrac{-352}{-6.4} = 55.$

To find *k*, substitute h = 55 in place of each *x* in the function to get:
$$k = P(55) = -3.2(55)^2 + 352(55) - 3600 = -3.2(3025) + 19,360 - 3600 = 6480.$$

The vertex is (55, 6480), the maximum point of the function. The x-coordinate represents the number of cattle. Since y = P(x), the y-coordinate represents the profit.

Answers: Based on the vertex, 55 cattle should be grazed on the pasture if the farmer wants to get the maximum profit of $6,480.

EXAMPLE 8   Finding the Minimum Value: At Starr Furniture Company, the **unit cost** in dollars to manufacture **x** tables is given by $C(x) = 0.004x^2 - 3x + 582$. For what number of tables is the unit cost at a minimum? What is the minimum unit cost?

Answers:

$C(x) = 0.004x^2 - 3x + 582$ is a second-degree function whose graph is a parabola. The leading coefficient is a = 0.004 which indicates that the parabola opens upward. The minimum point of the function is the vertex. To determine the minimum unit cost, find the vertex.

For this function, a = 0.004, b = -3, and c = 582.

The vertex = (h, k) where $h = \dfrac{-b}{2a} = \dfrac{-(-3)}{2(0.004)} = \dfrac{3}{0.008} = 375.$

To find k, substitute h = 375 in place of each x in the function to get:
$$k = C(375) = 0.004(375)^2 - 3(375) + 582 = 19.50.$$

The vertex is (375, 19.50), the minimum point of the function. The x-coordinate represents the number tables. Since y = C(x), the y-coordinate represents the unit cost in dollars.

Answers: Based on the vertex, the unit cost is at a minimum when 375 tables are manufactured. The minimum unit cost is $19.50 per table.

6.6   POLYNOMIAL AND RATIONAL INEQUALITIES

In Chapter 1 (Section 1.4) we observed that the linear inequality has no exponents in the expression. Polynomial and rational inequalities are called **non-linear inequalities**. The inequality expression contains an exponent, or, a fraction with a variable in the denominator. To solve a non-linear inequality, we will use the **Sign Chart Method**. As part of the process, we will substitute a "test number" in place of the variable and determine the *sign* of the final answer.

POLYNOMIAL INEQUALITIES

Given a polynomial inequality, a **critical value** is a number substituted for the variable which makes the polynomial *equal to zero*. (Use an equal sign whenever you state a critical value.) The critical values are not the solution set for the inequality. The critical values help us determine the solution set.

EXAMPLE 1   Find the critical value(s) for the polynomial inequality: $x^2 - 4x < 0$

Answer:

Factor the polynomial. $x(x - 4) < 0$

The critical values are: $x = 0$ and $x = 4$

These are the numbers which make $x^2 - 4x$ equal to zero.
(Check: $0^2 - 4(0) = 0 - 0 = 0$ and $4^2 - 4(4) = 16 - 16 = 0$)

Solving a Polynomial Inequality

As part of the process for solving a polynomial inequality, make a Sign Chart: Substitute one or more "test numbers" in place of the variable and determine the *sign* of the final answer. Also, graph the solution set on a standard number line and write the final answer in interval notation.

Polynomial Inequalities – The Sign Chart Method

- The right side of the inequality *must be zero*.
- *Factor* the polynomial completely.
- Find the *critical values* of the polynomial.
- Set up a *Sign Chart* which includes a number line. On the number line, mark the critical value(s): If the inequality has > or <, then use an open circle; if the inequality has ≥ or ≤, then use a closed point.

- For each interval of the number line, substitute a "test number" and determine the <u>*sign*</u> *of the polynomial for the interval*.

- *Graph* the solution set on a standard number line -- shade the interval(s) which make the inequality true: > 0 (or ≥ 0) means shade the <u>positive</u> results; < 0 (or ≤ 0) means shade the <u>negative</u> results.

- Write the solution set in *interval notation*.

(6.6)

EXAMPLE 2   Solve the inequality: $x^3 + x^2 > 6x$

Answer:

(a) The right side of the inequality must be zero.

$$x^3 + x^2 - 6x > 0$$

(b) Factor the polynomial completely.

$$x(x^2 + x - 6) > 0$$
$$x(x + 3)(x - 2) < 0$$

(c) Find the critical values of the polynomial.   Critical Values: $x = 0, -3, 2$

(d) Set up a Sign Chart: List the headings "Test #" and   ">" $\Rightarrow$ Place an open circle
"Signs" (on the left side). On the next line, to the right,   at $-3, 0,$ and 2.
draw a number line and mark the critical value(s):

For $>$ or $<$, use an open circle. For $\geq$ or $\leq$, use a closed point.

Sign Chart <u>Set-up</u> for the Inequality $x(x + 3)(x - 2) > 0$:

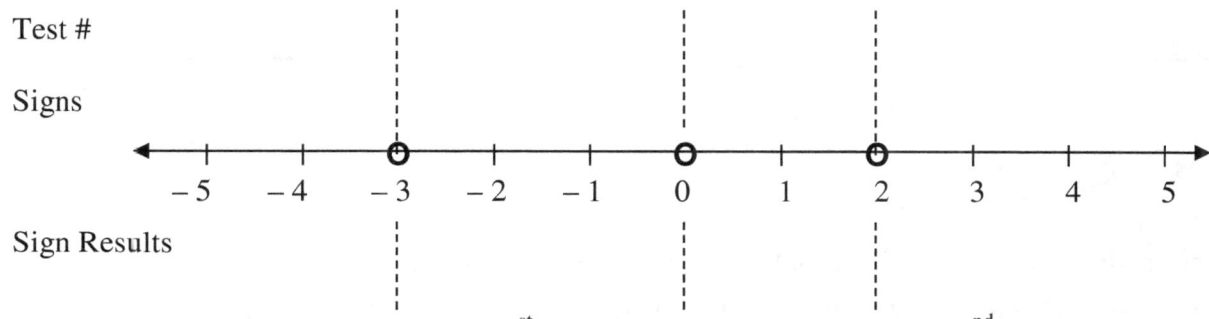

The number line now has 4 intervals. 1st interval: numbers less than -3; 2nd interval: numbers between -3 and 0; 3rd interval: numbers between 0 and 2; 4th interval: numbers greater than 2.

(e) For each interval, determine the sign of the polynomial for the interval: Choose any number in the interval as a "test number". Substitute the "test number" in place of x and give the <u>sign</u> of the polynomial for the interval. Write the resulting sign for each interval below the number line.

Sign Chart for the Inequality $\mathbf{x(x + 3)(x - 2) > 0}$:

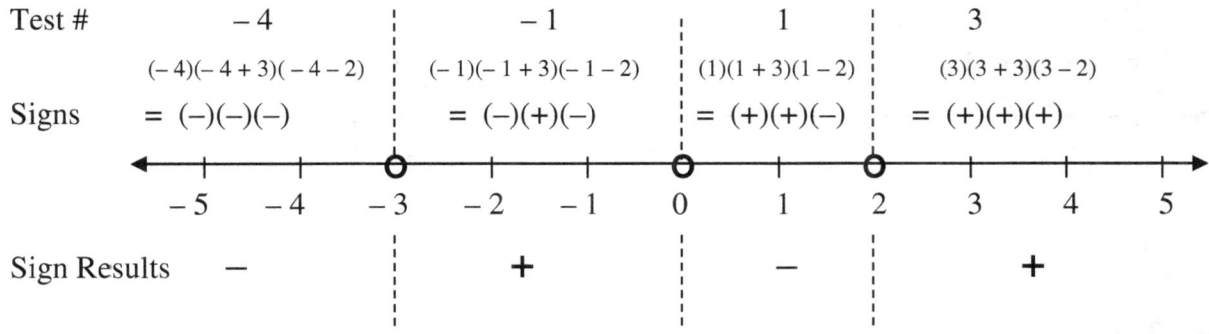

(f) Graph the solution set on a standard number line: Shade the interval(s) which make the inequality true: >0 means shade the <u>positive</u> results. Shade the 2nd and the 4th intervals.

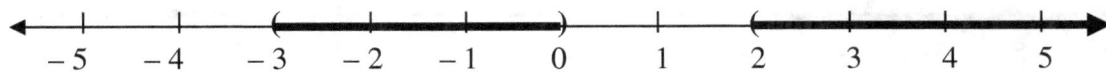

(g) Write the solution set in interval notation.   Solution: $(-3, 0) \cup (2, \infty)$

135

(6.6)

EXAMPLE 3   <u>A Polynomial Which Cannot Be Factored</u>:   Solve the inequality: $x^2 - 7 \geq 0$

Answer:

(a) The right side of the inequality is already zero.

$x^2 - 7 \geq 0$

(b) The binomial $x^2 - 7$ cannot be factored.
    To find the Critical Values, use the Square Root Method.

Critical Values: $x^2 = 7$
$$\sqrt{x^2} = \pm\sqrt{7}$$

(c) State the critical values of the polynomial as decimals.

$x = \sqrt{7} \approx 2.6,\ x = -\sqrt{7} \approx -2.6$

(d) Set up a Sign Chart for inequality $x^2 - 7 \geq 0$.

"$\geq$" $\Rightarrow$ Place a closed point
at -2.6 and 2.6.

Test #

Signs

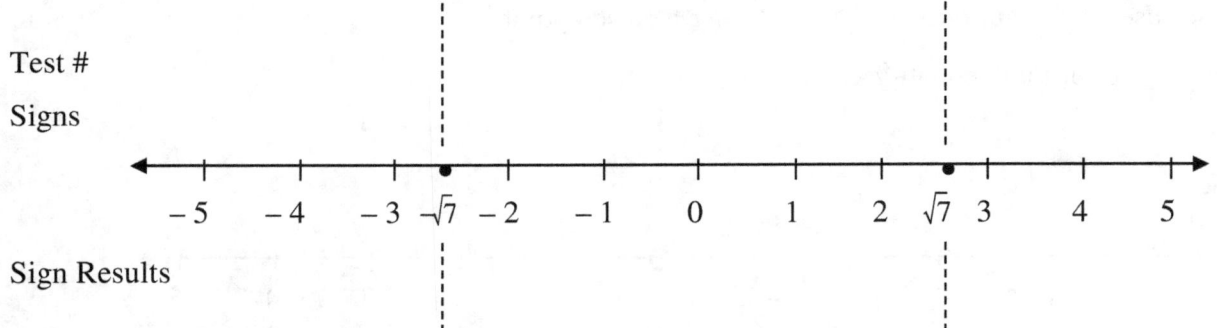

Sign Results

The number line now has 3 intervals. 1st interval: numbers less than $-\sqrt{7}$;

2nd interval: numbers between $-\sqrt{7}$ and $\sqrt{7}$; 3rd interval: numbers greater than $\sqrt{7}$.

(e) For each interval, determine the sign of the polynomial for the interval: Choose any number in the interval as a "test number". Substitute the "test number" in place of x and give the <u>sign</u> of the polynomial for the interval. Write the resulting sign for each interval below the number line.

Sign Chart for the Inequality $x^2 - 7 \geq 0$:

Test #	-3	0	3
	$(-3)^2 - 7$	$(0)^2 - 7$	$(3)^2 - 7$
Simplify	$= 9 - 7 = $ positive	$= 0 - 7 = $ negative	$= 9 - 7 = $ positive

Sign of Results $+$ $-$ $+$

(f) Graph the solution set on a standard number line: Shade the interval(s) which make the inequality true: ≥ 0 means shade the <u>positive</u> results. Shade the 1st and the 3rd intervals.

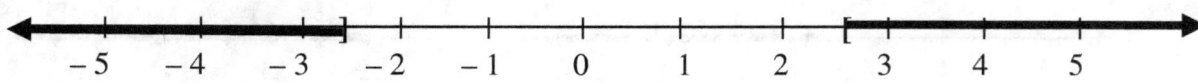

(g) Write the solution set in interval notation.   Solution: $(-\infty, -\sqrt{7}\,] \cup [\sqrt{7}, \infty)$

(6.6)

RATIONAL INEQUALITIES

A rational inequality involves a fraction which has a variable in the denominator. For a fraction, a **critical value** is a number which makes the numerator equal to zero. (Use an equal sign whenever you state a critical value.) A **restricted value** is a number which makes the denominator zero. Use a "$\neq$" sign to state a restricted value. (Remember that the denominator of a fraction cannot equal zero.)

EXAMPLE 4   Find the critical value(s) and the restricted value(s) for the rational inequality:

$$\frac{x^2 + 2x - 3}{x^2 - 4} \leq 0$$

Answer:

Factor the numerator and the denominator. $\dfrac{(x+3)(x-1)}{(x+2)(x-2)} \leq 0$

The critical values of the numerator: $x = -3$ and $x = 1$

The restricted values (of the denominator): $x \neq -2$ and $x \neq 2$

✖ *__Helpful Hints__ – Showing Critical Values and Restricted Values on the Number Line:*
(1) On the number line, a critical value may be open or closed, depending on the inequality symbol used in the problem. For $>$ or $<$, use an open circle. For $\geq$ or $\leq$, use a closed point.
(2) A restricted value *must be open on the number line*. (In other words, "$\neq$" indicates that the graph is open at the restricted value.)

For Example 4 above: "$\leq$" $\Rightarrow$ Place a closed point at the critical values -3 and 1.
        For the restricted values, "$\neq$" $\Rightarrow$ Place an open circle at -2 and 2.

Solving a Rational Inequality

The steps for solving a rational inequality are similar to the process for solving a polynomial inequality.

✖ *__Helpful Hint__* – **Important: When solving a rational inequality, _do not cancel out the denominator!_** Both the numerator and denominator are used to set up the Sign Chart.

EXAMPLE 5   Solve the inequality: $\dfrac{5x - 10}{3x - 3} \leq 0$

Answer:

(a) The right side of the inequality must be zero. $\dfrac{5x - 10}{3x - 3} \leq 0$

(b) Factor the numerator and denominator completely. $\dfrac{5(x - 2)}{3(x - 1)} \leq 0$

(c) Write the critical value of the numerator.   Critical Value: $x = 2$
                                   "$\leq$" $\Rightarrow$ Place a closed point at 2.

    Write the restricted value of the denominator.   Restricted Value: $x \neq 1$
                                   "$\neq$" $\Rightarrow$ Place an open circle at 1.

(6.6)

(d) Set up a Sign Chart. On the number line and mark the critical
value (closed point) and mark the restricted value (must be an open circle).

Test #

Signs

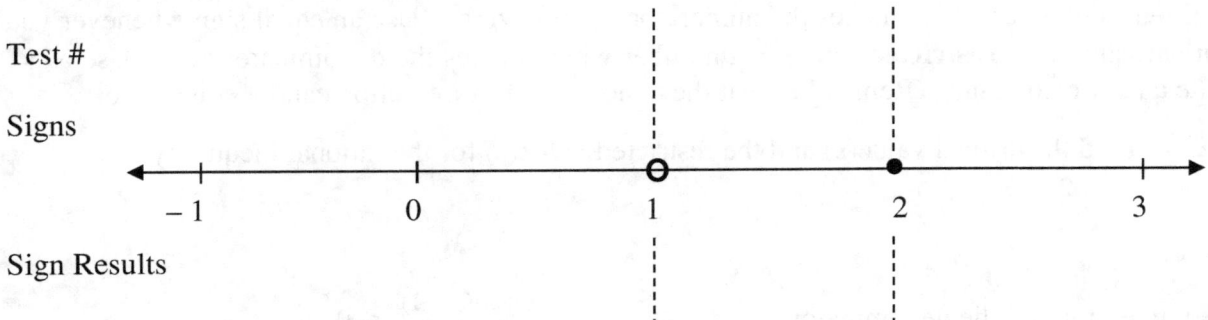

-1 0 1 2 3

Sign Results

The number line now has 3 intervals. 1st interval: numbers less than 1; 2nd interval: numbers between
1 and 2; 3rd interval: numbers greater than 2.

(e) For each interval, determine the sign of the fraction for the interval: Choose any number in the
interval as a "test number". Substitute the "test number" in place of x and give the <u>sign</u> of the fraction
for the interval. Write the resulting sign for each interval below the number line.

Sign Chart for the Inequality $\dfrac{5(x-2)}{3(x-1)} \le 0$:

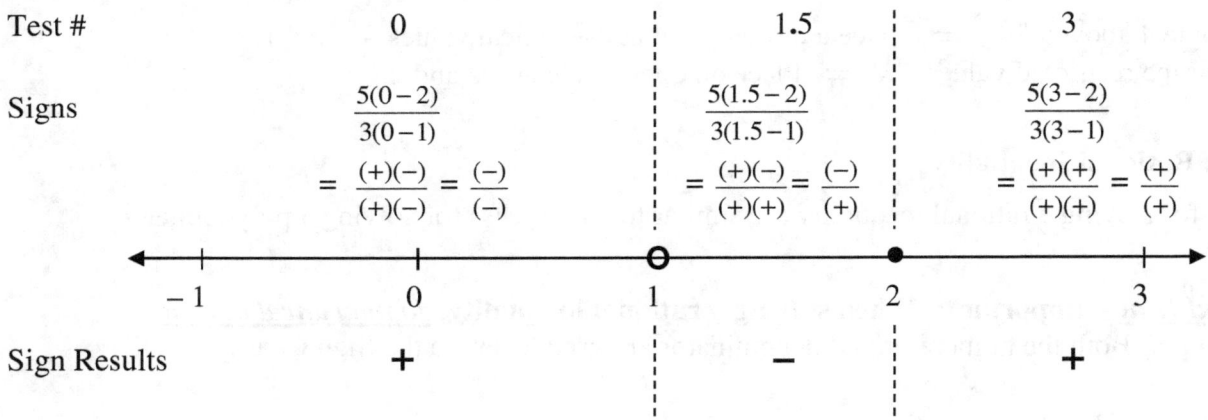

Test #	0	1.5	3
Signs	$\dfrac{5(0-2)}{3(0-1)}$	$\dfrac{5(1.5-2)}{3(1.5-1)}$	$\dfrac{5(3-2)}{3(3-1)}$
	$= \dfrac{(+)(-)}{(+)(-)} = \dfrac{(-)}{(-)}$	$= \dfrac{(+)(-)}{(+)(+)} = \dfrac{(-)}{(+)}$	$= \dfrac{(+)(+)}{(+)(+)} = \dfrac{(+)}{(+)}$

-1 0 1 2 3

Sign Results $+$ $-$ $+$

(f) Graph the solution set on a standard number line: Shade the interval(s) which make the inequality
true: ≤ 0 means shade the <u>negative</u> results. Shade the 2nd interval.

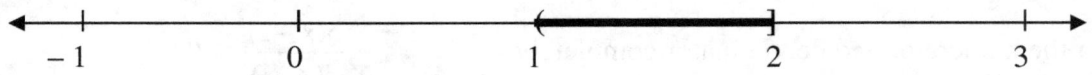

-1 0 1 2 3

(g) Write the solution set in interval notation.   Solution: $(1, 2]$

138

Chapter 7   Exponential and Logarithmic Functions

7.1   GRAPHING EXPONENTIAL FUNCTIONS (Including Base e)

Reminder--Scientific Calculators

This chapter requires the use of a scientific calculator. Each brand or model may have a different layout for the keypad. Use the appropriate special keys which appear on your calculator.

✳ *Helpful Hints* – (1) **The Exponent Key**: On your calculator keypad, you should have ONE of following keys: $[Y^X]$ or $[X^Y]$ or $[^\wedge]$. Whenever the *Mathco* study guide uses the $[Y^X]$ key, press the key which is on your calculator, $[Y^X]$ or $[X^Y]$ or $[^\wedge]$.

(2) **For Operations Written Above a Calculator Key**: On your calculator keypad, you should have ONE of following keys: [2nd] or [Shift] or [INV]. Whenever the *Mathco* study guide uses the [2nd] key, press the key which is on your calculator, [2nd] or [Shift] or [INV].

THE SCIENTIFIC CALCULATOR AND EXPONENTS

To compute 2^8 on your scientific calculator, press: 2 $[Y^X]$ $[(]$ 8 $[)]$ $[=]$.
Your screen should show: 256.

EXAMPLE 1   Use a scientific calculator to evaluate each expression.

(a) 5^4 Calculator: 5 $[Y^X]$ $[(]$ 4 $[)]$ $[=]$ Answer: 625

(b) 5^{-3} Calculator: 5 $[Y^X]$ $[(]$ $[(-)]$ 3 $[)]$ $[=]$ Answer: 0.008

(c) $7^{1.73205}$ Calculator: 7 $[Y^X]$ $[(]$ 1.73205 $[)]$ $[=]$ Answer: ≈ 29.09

EXPONENTIAL FUNCTIONS

The basic **exponential function** has the format $f(x) = b^x$, where $b > 0$ and $b \neq 1$. Observe that an exponential function has x in the exponent.

EXAMPLE 2   Let $f(x) = 5^x$ and $g(x) = \left(\dfrac{1}{2}\right)^x$. Evaluate each of the following.

(a) $f(3) = 5^3 = 125$

(b) $g(3) = \left(\dfrac{1}{2}\right)^3 = \dfrac{1^3}{2^3} = \dfrac{1}{8}$

(c) $f(0) = 5^0 = 1$

(d) $f\left(\dfrac{1}{2}\right) = 5^{\frac{1}{2}} = \sqrt{5} \approx 2.2361$

(e) $f(-3) = 5^{-3} = \dfrac{1}{5^3} = \dfrac{1}{125}$

(f) $g(-3) = \left(\dfrac{1}{2}\right)^{-3} = \left(\dfrac{2}{1}\right)^3 = 8$

(7.1)

The graph of a basic exponential function is an **exponential curve**. Two exponential functions are illustrated the following examples.

Given the exponential function $f(x) = b^x$ (b > 0 and b ≠ 1), the domain is all real numbers. For one end of the exponential curve, the y-coordinates are close to a horizontal boundary line called the **horizontal asymptote**. For the other end of the curve, the y-coordinates extend toward infinity.

EXAMPLE 3   Make a table of ordered pairs and graph $f(x) = 2^x$. Use the integers of the interval [– 3, 3] in the x column. Then find (a) the horizontal asymptote, (b) the domain, (c) the range, (d) the x-intercept, and (e) the y-intercept.

Answers:

x	$f(x) = 2^x$	(Workspace)
– 3	0.125	$2^{-3} = \dfrac{1}{2^3} = \dfrac{1}{8}$
– 2	0.25	$2^{-2} = \dfrac{1}{2^2} = \dfrac{1}{4}$
– 1	0.5	$2^{-1} = \dfrac{1}{2^1} = \dfrac{1}{2}$
0	1	$2^0 = 1$
1	2	$2^1 = 2$
2	4	$2^2 = 4$
3	8	$2^3 = 8$

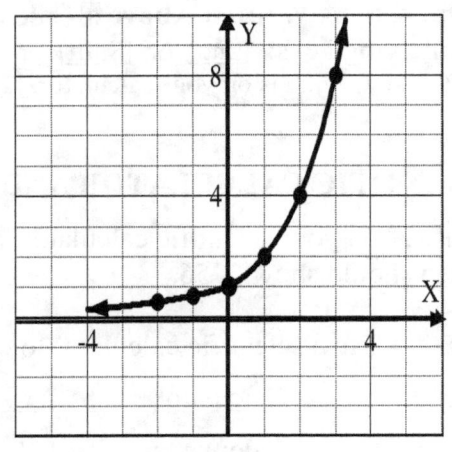

(a) The y-coordinates of $f(x) = 2^x$ are greater than zero. When a negative x-coordinate is less than – 1, the y-coordinate is close to zero. Therefore the horizontal asymptote is y = 0, which is the x-axis.

(b) The left end of the graph has an arrow which indicates – ∞. The right end of the graph has an arrow which indicates ∞. The domain for $f(x) = 2^x$ is {all real numbers} = (– ∞, ∞).

(c) As stated in answer (a), the y-coordinates of the graph are above zero. The range for $f(x) = 2^x$ is {y > 0} = (0, ∞).

(d) There is no x-intercept for f(x). The graph does not touch or cross the x-axis.

(e) The y-intercept is (0, 1).

(7.1)

✳ *__Helpful Hint__* – In the next example, the exponential function $g(x) = 2^{1-x}$ has a *negative coefficient for x*. The graph of g(x) is an exponential curve in which the high end of the curve is at the left. The lower end of the curve is at the right, close to the horizontal asymptote.

EXAMPLE 4   Make a table of ordered pairs and graph $g(x) = 2^{1-x}$. Use the integers of the interval [– 2, 4] in the x column. Then find (a) the horizontal asymptote, (b) the domain, (c) the range.

Answers:

x	$g(x) = 2^{1-x}$	(Workspace)
– 2	8	$2^{1-(-2)} = 2^3 = 8$
– 1	4	$2^{1-(-1)} = 2^2 = 4$
0	2	$2^{1-(0)} = 2^1 = 2$
1	1	$2^{1-(1)} = 2^0 = 1$
2	0.5	$2^{1-(2)} = 2^{-1} = \dfrac{1}{2^1} = \dfrac{1}{2}$
3	0.25	$2^{1-(3)} = 2^{-2} = \dfrac{1}{2^2} = \dfrac{1}{4}$
4	0.125	$2^{1-(4)} = 2^{-3} = \dfrac{1}{2^3} = \dfrac{1}{8}$

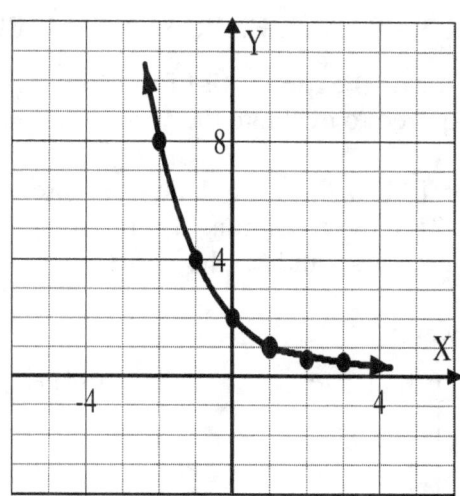

(a) The y-coordinates of $g(x) = 2^{1-x}$ are positive and not equal to zero. When a positive x-coordinate is greater than 2, the y-coordinate is close to zero. Therefore the horizontal asymptote is y = 0, which is the x-axis.

(b) The left end of the graph has an arrow which indicates – ∞. The right end of the graph has an arrow which indicates ∞. The domain for $g(x) = 2^{1-x}$ is {all real numbers} = (– ∞, ∞).

(c) As stated in answer (a), the y-coordinates of the graph are above zero. The range for $g(x) = 2^{1-x}$ is {y > 0} = (0, ∞).

THE IRRATIONAL NUMBER *e*

The number $e \approx$ **2.7182818284. . .** is an irrational number which is used as the base in some of the exponential formulas involving real-world phenomena, such as population growth, bacteria growth, continuous compound interest, and natural decay.

On the scientific calculator, the $[\,e^x\,]$ key is used to evaluate expressions involving the number *e*.

To get *e* on the calculator, press: [2nd] [e^x] [( ] 1 [ )]   [=].
The result is the decimal 2.7183 (rounded to four decimal places).

(7.1)

EXAMPLE 5   Use a scientific calculator to evaluate each expression, rounded to 4 decimal places.

(a) e^3 Calculator: [2nd] [e^x] [( ] 3 [ )] [=]   Answer: 20.0855

(b) $e^{-1.3}$ Calculator: [2nd] [e^x][( ] [(−)] 1.3 [ )] [=]   Answer: 0.2725

(c) $e^{2.1} - 5$ Calculator: [2nd] [e^x] [( ] 2.1 [ )] [−] 5 [=]   Answer: 3.1662

(d) $10e^2$ Calculator: 10 [×] [2nd] [e^x] [( ] 2 [ )] [=]   Answer: 73.8906

THE NATURAL EXPONENTIAL FUNCTION

The **natural exponential function** has the format $f(x) = e^x$, where $e \approx 2.7183$. A scientific calculator is required to evaluate the y-coordinates for the natural exponential function, rounded to a decimal.

EXAMPLE 6   Make a table of ordered pairs and graph $f(x) = e^x - 1$. Use the integers of the interval [− 3, 3] in the x column. Then find (a) the horizontal asymptote, (b) the domain, (c) the range, (d) the x-intercept(s), and (e) the y-intercept.

Answers:

x	$f(x) = e^x - 1$	(Workspace)
− 3	− 0.95	$e^{-3} - 1$
− 2	− 0.86	$e^{-2} - 1$
− 1	− 0.63	$e^{-1} - 1$
0	0	$e^0 - 1$
1	1.72	$e^1 - 1$
2	6.38	$e^2 - 1$
3	19.09	$e^3 - 1$

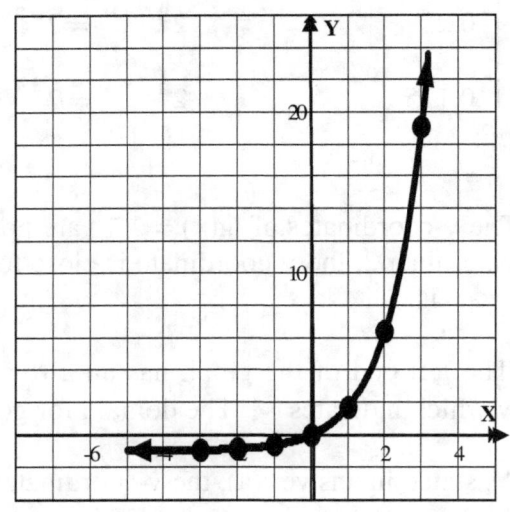

(a) The y-coordinates of $f(x) = f(x) = e^x - 1$ are greater than − 1. When a negative x-coordinate is less than − 1, the y-coordinate is close to − 1. Therefore the horizontal asymptote is y = − 1.

(b) The left end of the graph has an arrow which indicates − ∞. The right end of the graph has an arrow which indicates ∞. The domain for $f(x) = e^x - 1$ is {all real numbers} = (− ∞, ∞).

(c) As stated in answer (a), the y-coordinates of the graph are above − 1. The range for $f(x) = e^x - 1$ is {y > − 1} = (− 1, ∞).

(d) The x-intercept for f(x) is (0, 0).

(e) The y-intercept is (0, 0).

APPLICATIONS

Compound Interest Formula: $A = P\left(1+\frac{r}{n}\right)^{nt}$

The above compound interest formula is used for investment or savings account problems for which there is one deposit placed in an account for a period of time. It is also used for loan problems. When using the compound interest formula:

P = principal (the original deposit, or, the money borrowed)
r = interest rate (stated as a decimal)
n = number of times interest is paid per year
t = time (in years)
A = total amount (new balance) including interest earned

Helpful Hints – Interest Payments Per Year

If Interest is Compounded	Number of Interest Payments Per Year
annually	n = 1 interest payment
semi-annually	n = 2 interest payments
quarterly	n = 4 interest payments
monthly	n = 12 interest payments
weekly	n = 52 interest payments
daily	n = 365 interest payments

EXAMPLE 7 Investment problem: If $2,500 is invested at 6% compounded quarterly, how much money will the account have 5 years later?

Answer: Principal P = $2,500, interest rate r = 6% = 0.06, n = 4 interest payments per year, t = 5 years. A = the new balance at the end of 5 years. Using the compound interest formula,

$$A = 2500\left(1+\frac{0.06}{4}\right)^{(4)(5)} = 2500\left(1+\frac{0.06}{4}\right)^{20}$$

Using a scientific calculator and rounding to the nearest cent, A = $3,367.14 is the new balance.

Calculator: 2500 [x] [( ] 1 [+] .06 [÷] 4 [ )] [y^x] 20 [=]

EXAMPLE 8 Loan Problem: If $3,000 is borrowed at 9.5% compounded monthly, how much money will be due at the end of 4 years?

Answer: Principal P = $3,000, interest rate r = 9.5% = 0.095, n = 12 interest payments per year, t = 4 years, and A = the amount due at the end of 4 years. Using the compound interest formula,

$$A = 3000\left(1+\frac{0.095}{12}\right)^{(12)(4)} = 3000\left(1+\frac{0.095}{12}\right)^{48}$$

Using a scientific calculator and rounding to the nearest cent, A = $4,380.29.

(7.1)

Continuous Compound Interest Formula: $A = Pe^{rt}$

This formula gives the amount when the interest is compounded "every instant."

Helpful Hints – (1) The above formula is used only when the problem states "continuous" or "continuously." Do not interchange the original compound interest formula with the continuous compound interest formula. Compare Example 9 to Example 7. The answers are not the same. (2) When using the continuous compound interest formula, do not list *e* as a variable. Remember that $e \approx 2.7182818284\ldots$ is a constant. Use your scientific calculator to evaluate expressions involving *e*.

EXAMPLE 9 Investment problem: If $2,500 is invested at 6% compounded continuously, how much money will the account have 5 years later?

Answer: Principal P = $2,500, interest rate r = 6% = 0.06, t = 5 years and A = the new balance at the end of 5 years.

Use the continuous compound interest formula to write the equation. $A = 2500e^{(0.06)(5)}$

Simplify the exponent. $A = 2500e^{0.3}$

Use a scientific calculator to evaluate the expression, rounding to the nearest cent. A = $3,374.65 is the new balance.

Calculator: 2500 [x] [2nd] [e^x] .3 [=]

Present Value is the amount of money (the Principal) which must be invested today in order to produce a desired account balance at a later date.

EXAMPLE 10 Present Value: An investment which compounds continuously at 9.25% will be worth $10,000 at the end of 7 years. Find the present value of this investment to the nearest dollar.

Answer: Amount A = $10,000, interest rate r = 9.25% = 0.0925, t = 7 years and P = the principal or present value.

Use the continuous compound interest formula to write the equation. $10,000 = Pe^{(0.0925)(7)}$

Simplify the exponent. $10,000 = Pe^{0.6475}$

Solve for *P*. $$\frac{10,000}{e^{0.6475}} = \frac{Pe^{0.6475}}{e^{0.6475}}$$

$$P = \frac{10,000}{e^{0.6475}}$$

Use a scientific calculator to evaluate the expression, rounding to the nearest dollar. $P \approx \$5,234$

The present value of $5,234 must be invested today in order to produce $10,000 at the end of 7 years.

Population Growth   Formula: $P = P_0 e^{rt}$ where P_0 = original population,

r = rate of growth, t = time in years and

P = the resulting population

EXAMPLE 11 If in a certain year the world population was about 6.85 billion people and the population grew continuously at an annual rate of 1.5%, what would be the estimated world population 6 years later? (Round the answer to two significant digits.)

Answer:

P_0 = 6.85 billion, growth rate r = 1.5% = 0.015, t = 6 years and P = the population at the end of 6 years.

Use the population growth formula
  to write the equation. $P = 6.85e^{(0.015)(6)}$

Simplify the exponent. $P = 6.85e^{0.09}$

Use a scientific calculator to evaluate the
  expression, rounding to two significant digits. $P \approx 7.5$ billion

7.2   PROPERTIES OF LOGARITHMS

DEFINITION OF LOGARITHM

A **logarithm** means, "What is the *exponent* if we are given the base and the original result?"
Given the **exponential form** $2^3 = 8$, the **logarithmic form** is:

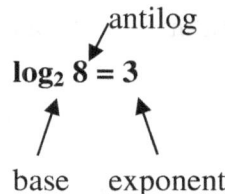

The original result "8" is called the **antilog**.

The base and the antilog must be <u>positive</u> real numbers.

The answer to the logarithm can be any real number.

The expression $\log_2 8 = 3$ is read, "log base 2 of 8 equals 3." The expression $\log_2 8$ means: "2 to what exponent power equals 8?" The exponent is 3.

Observe that the answer to a logarithm is the exponent. Therefore: $\boldsymbol{\log_B A = X \iff B^X = A}$

EXAMPLE 1   Re-write each expression from exponential form to logarithmic form.

(a) $2^3 = 8$ Answer: $\log_2 8 = 3$ (b) $5^2 = 25$ Answer: $\log_5 25 = 2$

(c) $3.5^7 = N$ Answer: $\log_{3.5} N = 7$ (d) $\sqrt{16} = 16^{\frac{1}{2}} = 4$ Answer: $\log_{16} 4 = \dfrac{1}{2}$

EXAMPLE 2   Re-write each expression from logarithmic form to exponential form.

(a) $\log_3 9 = 2$ Answer: $3^2 = 9$ (b) $\log_2 128 = 7$ Answer: $2^7 = 128$

(c) $\log_n p = k$ Answer: $n^k = p$ (d) $\log_{10} 0.001 = -3$ Answer: $10^{-3} = 0.001$

(7.2)

TWO SPECIAL LOGARITHMS

There are two logarithms which are programmed into your calculator.

Common logarithm: $\log x = \log_{10} x$ **natural logarithm:** $\ln x = \ln_e x$

✖ *__Helpful Hints__* – (1) A logarithm written without a base is always base 10. It is called a common logarithm. (2) The natural logarithm is always base *e*. (3) Do not interchange a common logarithm with a natural logarithm. (4) All other logarithms must have the base given with the expression.

EXAMPLE 3   Rewrite each logarithm in exponential form:

(a) $\log x = 15$ Answer: $10^{15} = x$ (b) $\log R = T$ Answer: $10^T = R$

(c) $\ln x = 15$ Answer: $e^{15} = x$ (d) $\ln R = T$ Answer: $e^T = R$

Rewrite each exponential expression in logarithmic form:

(e) $10^2 = 100$ Answer: $\log 100 = 2$ (f) $e^4 = 54.5982$ Answer: $\ln 54.5982 = 4$

PROPERTIES OF LOGARITHMS

For each property, base *b* is a positive real number.

PROPERTY 1: $\log_b a = x \iff b^x = a$

PROPERTY 2: $\log_b b = 1$ (If the base and the antilog are the same, the logarithm cancels out to 1.)

PROPERTY 3: $\log_b b^x = x$ (The base and the antilog are the same; the logarithm cancels out.)

PROPERTY 4: $\log_b 1 = 0$

       EXAMPLE 4   Evaluate each logarithm:

         (a) $\log 10 = \log_{10} 10 = 1$ (b) $\ln e = \ln_e e = 1$ (c) $\log_7 7^3 = 3$

PROPERTY 5: $\log_b xy = \log_b x + \log_b y$

       EXAMPLE 5 $\log_2 5v = \log_2 5 + \log_2 v$

> **PRODUCT RULE**
> Add the separate logarithms.

PROPERTY 6: $\log_b \dfrac{x}{y} = \log_b x - \log_b y$

         EXAMPLE 6 $\log_2 \dfrac{5}{v} = \log_2 5 - \log_2 v$

> **QUOTIENT RULE**
> Subtract the separate logarithms.

PROPERTY 7: $\log_b x^r = r \log_b x$

         EXAMPLE 7 $\log_2 v^3 = 3 \log_2 v$

> **EXPONENT RULE**
> Multiply the exponent times the logarithm.

EXAMPLE 8   Use the Product Rule (Property 5) to rewrite: $\log_b w(w + 2)$

Answer: $\log_b w(w + 2) = \log_b w + \log_b (w + 2)$

�֍ **_Helpful Hint_ – Observe the second term: For the antilog, you cannot "split" a polynomial.**

EXAMPLE 9   Use the Product Rule and the Quotient Rule (Properties 5 and 6) to rewrite $\log_b \dfrac{uv}{wx}$.

Shortcut Answer: $\log_b \dfrac{uv}{wx} = \log_b u + \log_b v - \log_b w - \log_b x$

✖ **_Helpful Hints_ – In Example 9 above: (1) For each factor of the numerator, add the separate logarithms. (2) For each factor of the denominator, subtract the separate logarithms.**

EXAMPLE 10   Rewrite the logarithm in expanded form using every property that applies to the expression: $\log_b \dfrac{u^3 v^5}{\sqrt{w}}$.

Answer:

(a) Write the denominator in exponent form.

$$\log_b \frac{u^3 v^5}{\sqrt{w}} = \log_b \frac{u^3 v^5}{w^{\frac{1}{2}}}$$

(b) Use the Product Rule to write the numerator as a sum of the separate logarithms and use the Quotient Rule to write the denominator as the subtraction of the logarithm.

$$= \log_b u^3 + \log_b v^5 - \log_b w^{\frac{1}{2}}$$

(c) Use the Exponent Rule to multiply each exponent times its logarithm.

$$= 3 \log_b u + 5 \log_b v - \frac{1}{2} \log_b w$$

THREE PROPERTIES IN REVERSE

Write each expression as ONE logarithm.

PROPERTY 7   Exponent Rule in reverse: $r \log_b m = \log_b m^r$
                   (Change the coefficient r to an exponent.)

    EXAMPLE 11 $5 \log_3 u = \log_3 u^5$

PROPERTY 6   Quotient Rule in reverse: $\log_b m - \log_b n = \log_b \dfrac{m}{n}$

               (Change the difference of logarithms to _one_ logarithm of a quotient.)

    EXAMPLE 12 $2 \log_3 w - 7 \log_3 x = \log_3 w^2 - \log_3 x^7 = \log_3 \dfrac{w^2}{x^7}$

PROPERTY 5   Product Rule in reverse: $\log_b m + \log_b n = \log_b mn$

               (Change the sum of logarithms to _one_ logarithm of a product.)

    EXAMPLE 13 $5 \log_3 u + 4 \log_3 v = \log_3 u^5 + \log_3 v^4 = \log_3 u^5 v^4$

(7.2)

EXAMPLE 14   Rewrite the logarithmic expression as *one* logarithm:
$$2 \log_6 w + 3 \log_6 x - 4 \log_6 p - \log_6 r$$

Answer:

Use the Exponent Rule in reverse to change each coefficient to an exponent.

$$2 \log_6 w + 3 \log_6 x - 4 \log_6 p - \log_6 r = \log_6 w^2 + \log_6 x^3 - \log_6 p^4 - \log_6 r$$

Product Rule in reverse: If the sign in front of the logarithm is positive, write the antilog in the numerator. Quotient Rule in reverse: If the sign in front of the logarithm is negative, write the antilog in the denominator.

$$\log_6 w^2 + \log_6 x^3 - \log_6 p^4 - \log_6 r = \log_6 \frac{w^2 x^3}{p^4 r}$$

7.3   SOLVING LOGARITHMIC EQUATIONS

Steps:
- Write the logarithm(s) on the left side of the equation.
- The equation must be in the format: *One logarithm = a real number*
- Change equation to *exponential form*.
- *Solve* for x.
- *Check* that the final answer makes the antilog of the original equation a positive real number.

EXAMPLE 1   Solve for x: $\log_3 x = 5$

The logarithm is on the left side of the equation. $\log_3 x = 5$

Write the equation in exponential form. $3^5 = x$

Simplify the result.   Answer: x = 243

The result 243 is positive, which is acceptable as the antilog
   of the original equation.

EXAMPLE 2   Solve for x; round the answer to two decimal places: $\log x = 0.75$

Answer:

The logarithm is on the left side of the equation. $\log x = 0.75$
The base for a common logarithm is 10. $\log_{10} x = 0.75$

Write the equation in exponential form. $10^{0.75} = x$

Use a scientific calculator to find the result.   Answer: x = 5.62

The result 5.62 is positive, which is acceptable as the antilog
   of the original equation.

EXAMPLE 3   Solve for x; round the answer to four decimal places: $\ln x = -2.5$

The logarithm is on the left side of the equation.

$\ln x = -2.5$

The base for a natural logarithm is e.

$\ln_e x = -2.5$

Write the equation in exponential form.

$e^{-2.5} = x$

Use a scientific calculator to find the result.

Answer: $x = 0.0821$

The result 0.0821 is positive, which is acceptable as the antilog of the original equation.

EXAMPLE 4   Solve for x: $\log_3 20 + \log_3 x - \log_3 5 = 4$

Use the Product Rule and Quotient Rule in reverse to write the left side of the equation as one logarithm.

$$\log_3 \frac{20x}{5} = 4$$

Reduce the fraction.

$\log_3 4x = 4$

Write the equation in exponential form.

$3^4 = 4x$

Simplify the result.

$81 = 4x$

Solve the equation for x.

Answer: $x = 20.25$

The result 20.25 is positive, which is acceptable as the antilog of the original equation.

EXAMPLE 5   Solve for x: $\log (x - 3) + \log x = 1$

Use the Product Rule in reverse to write the left side of the equation as one logarithm.

$\log (x - 3)(x) = 1$

The base for a common logarithm is 10.

$\log_{10} (x - 3)(x) = 1$

Write the equation in exponential form.

$10^1 = (x - 3)(x)$

Simplify the result.

$10 = x^2 - 3x$

The equation has an exponent. Cancel the left side of the equation.

$0 = x^2 - 3x - 10$

Factor the trinomial.

$0 = (x - 5)(x + 2)$

Then solve for x.

Possible results: $x = 5$ or $x = -2$

The final answer(s) must make each antilog positive:

If $x = 5$, then the antilog $x - 3 = 5 - 3 = 2$ which is positive.

If $x = -2$, then the antilog $x - 3 = -2 - 3 = -5$ which is negative.

Therefore $x \neq -2$.

Answer: $x = 5$

(7.3)

✴ *Helpful Hint* – If an equation has a logarithm on both sides of the equation, rewrite the equation with the logarithms on the left side and the constant term on the right side of the equation. Then use the steps for solving a logarithmic equation. The process is shown in the following example.

EXAMPLE 6   Solve $4 \ln 2 = \ln(x + 7)$ for x.

Write the logarithms on the left side of the equation.

$$4 \ln 2 - \ln(x + 7) = 0$$

Use the Exponent Rule in reverse to change the coefficient 4 to an exponent.

$$\ln 2^4 - \ln(x + 7) = 0$$

Then use the Quotient Rule in reverse to re-write the left side as one logarithm.

$$\ln \frac{2^4}{x+7} = 0$$

The base for a natural logarithm is *e*.

$$\ln_e \frac{2^4}{x+7} = 0$$

Write the equation in exponential form.

$$e^0 = \frac{2^4}{x+7}$$

Simplify the equation.

$$1 = \frac{16}{x+7}$$

To eliminate the fraction, multiply the LCD "x + 7" times both sides.

$$(x+7)1 = \frac{16}{x+7}(x+7)$$

Simplify the equation.

$$x + 7 = 16$$

Solve the equation for x.

Answer: $x = 11$

The result 11 is positive, which is acceptable as part of the antilog of the right side of the original equation.

7.4   GRAPHING LOGARITHMIC FUNCTIONS

COMMON AND NATURAL LOGARITHMS

To evaluate a common logarithm or a natural logarithm, use a scientific calculator.

EXAMPLE 1   Evaluate each logarithmic expression. Round the decimal answers to 4 decimal places.

(a) $\log 0.01$ Calculator: [log] .01 [=]   Answer: -2

(b) $\log 2.3$ Calculator: [log] 2.3 [=]   Answer: 0.3617

(c) $\ln 0.01$ Calculator: [ln] .01 [=]   Answer: -4.6052

(d) $\ln 2.3$ Calculator: [ln] 2.3 [=]   Answer: 0.8329

(e) $2 \ln 3$ Calculator: 2 [x] [ln] 3 [=]   Answer: 2.1972

(7.4)

OTHER LOGARITHMS

To evaluate logarithms to other bases, use the formula:

$$\log_B A = \frac{\log A}{\log B}$$

EXAMPLE 2   Evaluate the logarithmic expression $\log_5 234$.
Round the answers to 4 decimal places.

Answer: $\log_5 234 = \dfrac{\log 234}{\log 5} = 3.3896$

Calculator: [( ] [log] 234 [ )] [÷] [log] [( ] 5 [ )] [=]

GRAPHING A LOGARITHMIC FUNCTION

The graph of a basic logarithmic function is a **logarithmic curve,** as illustrated in the examples that follow.

For the **common logarithmic function** $f(x) = \log x$, the domain is $\{x > 0\}$; the x-coordinates (antilogs) must be positive. At the left end of the logarithmic curve, the points are very close to a vertical boundary line called the **vertical asymptote**. For the right end of the curve, the y-coordinates gradually increase, extending toward infinity.

✳ _**Helpful Hint**_ – To set up the table of ordered pairs for $f(x) = \log x$, observe that zero cannot be used in the x-column. Start the x-coordinates with a small decimals between 0 and 1, such as 0.01, 0.1, 0.3, and 0.5; then use the integers 1, 2, 3, etc. Use your calculator to find the y-coordinates.

EXAMPLE 3   (a) State the domain and make a table of ordered pairs and graph $f(x) = \log x$.
Then find: (b) the vertical asymptote, (c) the range, (d) the x-intercept, and (e) the y-intercept.

Answers:

(a) The domain of $f(x)$ is $x > 0$ since the antilogs must be positive.

x	f(x) = log x	(Workspace)
0.01	– 2	log 0.01
0.1	– 1	log 0.1
0.3	– 0.5	log 0.3
0.5	– 0.3	log 0.5
1	0	log 1
2	0.3	log 2
3	0.5	log 3
4	0.6	log 4
5	0.7	log 5

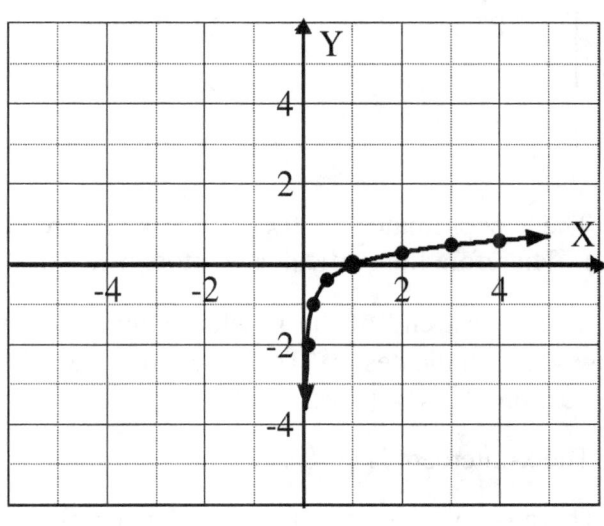

(7.4)

(b) The x-coordinates of f(x) = log x are greater than zero (i.e. the x-coordinates are to the right of zero). Therefore the vertical asymptote is x = 0, which is the y-axis.

(c) The lowest end of the graph has an arrow indicating –∞. The highest end of the graph has an arrow which indicates ∞. Thus y-coordinates of the graph are from –∞ to ∞ and the range for f(x) is {All real numbers}= (– ∞, ∞).

(d) The x-intercept is (1, 0).

(e) There is no y-intercept for f(x). The graph does not cross the y-axis.

For the **natural logarithmic function**, we use a similar procedure.

EXAMPLE 4   (a) State the domain and make a table of ordered pairs and graph $f(x) = \ln x^2$.
Then find: (b) the vertical asymptote, (c) the range, (d) the x-intercept, and (e) the y-intercept.

Answers:

Use the Exponent Rule for logarithms to rewrite $f(x) = \ln x^2$ as $f(x) = 2 \ln x$.

(a) The domain of f(x) is x > 0 since the antilogs must be positive.

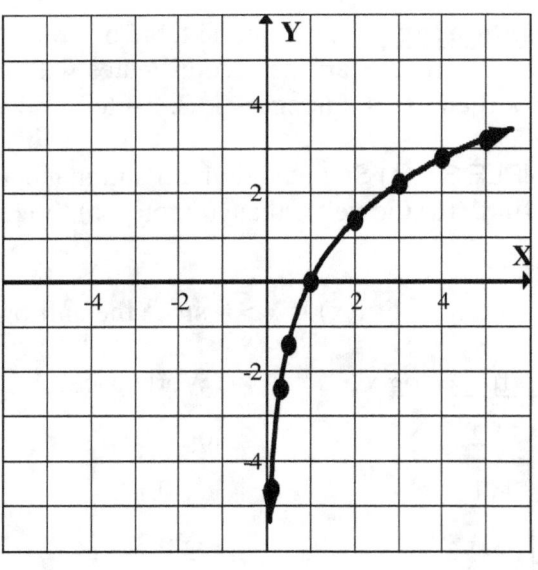

x	$f(x) = \ln x^2 = 2 \ln x$	(Workspace)
0.1	– 4.6	2 ln 0.1
0.3	– 2.4	2 ln 0.3
0.5	– 1.4	2 ln 0.5
1	0	2 ln 1
2	1.4	2 ln 2
3	2.2	2 ln 3
4	2.8	2 ln 4
5	3.2	2 ln 5

(b) The x-coordinates of f(x) = 2 ln x are greater than zero (i.e. the x-coordinates are to the right of zero). Therefore the vertical asymptote is x = 0, which is the y-axis.

(c) The lowest end of the graph has an arrow indicating –∞. The highest end of the graph has an arrow which indicates ∞. Thus y-coordinates of the graph are from – ∞ to ∞ and the range for f(x) is {All real numbers}= (– ∞, ∞).

(d) The x-intercept is (1, 0).

(e) There is no y-intercept for f(x). The graph does not cross the y-axis.

7.5 SOLVING EXPONENTIAL EQUATIONS

REVIEW – THE EXPONENT RULE FOR LOGARITHMS

$\log_b x^r = r \log_b x$ (Multiply the exponent times the logarithm.)

EXAMPLE 1 Rewrite each logarithm without using an exponent.

(a) $\log_b x^3 = 3 \log_b x$ (b) $\log 7^{3x} = 3x \log 7$

(c) $\ln 5^{2x+1} = (2x + 1)\ln 5$ *Use parentheses for multiplication of a polynomial times a logarithm.*

SOLVING EXPONENTIAL EQUATIONS

An exponential equation has a variable in the exponent.

Helpful Hint - *Whenever the variable is in the exponent of an equation,* ***use a logarithm on both sides of the equation.*** Then use the properties of logarithms and solve the equation.

Steps:

- Use *"log" on both sides* of the equation. If the base is *e*, *use "ln" on both sides* of the equal sign.
- Use the *Exponent Rule*: Multiply the exponent times the logarithm.
- Solve the equation for x and *write the exact logarithmic answer.*
- Then use your calculator to *find the decimal answer.*

EXAMPLE 2 Solve for x: $10^x = 14.3$

Answer:	$10^x = 14.3$
Use *log* on both sides of the equation.	$\log 10^x = \log 14.3$
Use the Exponent Rule to multiply the exponent times the logarithm.	$x \log 10 = \log 14.3$
The base for the common logarithm is 10. For this equation, the base and the antilog are the same. Cancel out the logarithm on the left side of the equation.	$x \cancel{\log_{10} 10} = \log 14.3$
Solve for *x* and write the exact logarithmic answer.	$x = \log 14.3$
Use your calculator to find the decimal answer. (Round the result to 4 significant digits.)	$x = 1.155$

EXAMPLE 3 Solve for x: $7^{3x} = 20$

Answer:	$7^{3x} = 20$
Use *log* on both sides of the equation.	$\log 7^{3x} = \log 20$
Use the Exponent Rule to multiply the exponent times the logarithm.	$3x \log 7 = \log 20$

(7.5)

Divide both sides of the equation by 3 log 7.

$$\frac{\cancel{3x \log 7}}{\cancel{3 \log 7}} = \frac{\log 20}{3 \log 7}$$

Solve for x and write the exact logarithmic answer.

$$x = \frac{\log 20}{3 \log 7}$$

Use your calculator to find the decimal answer. Round it to 4 decimal places.
   Calculator: $\underline{[\log]}\,\underline{[(}\,20\,\underline{)]}\,\underline{[\div]}\,\underline{[(}\,3\,\underline{[x]}\,\underline{[\log]}\,7\,\underline{)]}\,\underline{[=]}$

$$x = 0.5132$$

EXAMPLE 4   Solve for x: $e^{2x+3} = 5$

Answer:

$$e^{2x+3} = 11$$

The base is e. Use *ln* on both sides of the equation.

$$\ln e^{2x+3} = \ln 11$$

Use the Exponent Rule to multiply the exponent times the logarithm.

$$(2x+3)\ln e = \ln 11$$

The base for the natural logarithm is e.
   For this equation, the base and the antilog are the same.
   Cancel out the logarithm on the left side of the equation.

$$(2x+3)\cancel{\ln e} = \ln 11$$

Solve for x and write the exact logarithmic answer.
   Subtract 3 on both sides of the equation.

$$2x + 3 = \ln 11$$
$$\underline{-3 \quad -3}$$

$$2x = -3 + \ln 11$$

Divide both sides of the equation by 2.

$$x = \frac{-3 + \ln 11}{2}$$

Use your calculator to find the decimal answer. Round it to 3 decimal places.
   Calculator: $\underline{[(}\,\underline{[(-)]}\,3\,\underline{[+]}\,\underline{[\ln]}\,\underline{[(}\,11\,\underline{)]}\,\underline{)]}\,\underline{[\div]}\,2\,\underline{[=]}$

$$x = -0.301$$

APPLICATIONS

Exponential (Uninhibited) Growth   Formula: $P = P_0 e^{rt}$

(Observe that the exponent in the growth formula is *positive*.)

EXAMPLE 5   How many years did it take for the world population to grow from 4 billion people to 5 billion if it grows continuously at an annual rate of 2%?

Answer:

Given the initial population $P_0 = 4$ billion, the ending population $P = 5$ billion, and $r = 2\% = 0.02$, we will use the growth formula to find the time t in years.

Substitute the numbers into the formula.

$$5 = 4e^{0.02t}$$

Divide by 4 on both sides of the equation.

$$1.25 = e^{0.02t}$$

The base is e. Use *ln* on both sides of the equation.

$$\ln 1.25 = \ln e^{0.02t}$$

154

Use the Exponent Rule to multiply the exponent times the logarithm. $\ln 1.25 = (0.02t)\ln e$

Cancel out the logarithm on the right side of the equation. $\ln 1.25 = 0.02t$

Solve for t and write the exact logarithmic answer. $t = \dfrac{\ln 1.25}{0.02}$

Use your calculator to find the decimal answer. Round it to the nearest year. $t = 11$ years

Continuous Compound Interest – Finding the Interest Rate   Formula: $A = Pe^{rt}$

EXAMPLE 6   Vivian invested $1,600 which increased to $2,500 in five years. Find the annual interest rate if the interest compounded continuously.

Answer:

The principal is P = $1,600, the ending balance is A = $2,500, and time t = 5 years. We will use the continuous compound interest formula to find the annual interest rate r.

Substitute the numbers into the formula. $2,500 = 1,600e^{(r)(5)}$

Divide by 1,600 on both sides of the equation. $1.5625 = e^{5r}$

The base is e. Use ln on both sides of the equation. $\ln 1.5625 = \ln e^{5r}$

Use the Exponent Rule to multiply the exponent times the logarithm. $\ln 1.5625 = (5r)\ln e$

Cancel out the logarithm on the right side of the equation. $\ln 1.5625 = 5r$

Solve for r and write the exact logarithmic answer. $r = \dfrac{\ln 1.5625}{5}$

Use your calculator to find the decimal answer, rounded to three decimal places. $r = 0.089$

Write the annual interest rate as a percent. $r = 8.9\%$

✳ ***Helpful Hint*** – For all applications: If the principal (or initial value) is not given, use 1 for the initial value. Read the example below.

Continuous Compound Interest – Doubling Time

EXAMPLE 7   How many years, to the nearest year, will it take an investment to double in value if the annual interest rate is 10% compounded continuously?

Answer:

The initial investment amount is not given. Therefore let the principal P = $1. Since the investment will *double* at the end of t years, the ending balance will be A = $2. The interest rate r = 10% = 0.1. We will use the continuous compound interest formula to find the doubling time t to the nearest year.

Substitute the numbers into the formula $A = Pe^{rt}$. $2 = 1e^{(0.1)(t)}$

(7.5)

Simplify the exponent on the right side of the equation.	$2 = e^{0.1t}$
The base is *e*. Use *ln* on both sides of the equation.	$\ln 2 = \ln e^{0.1t}$
Use the Exponent Rule to multiply the exponent times the logarithm.	$\ln 2 = (0.1t)\ln e$
Cancel out the logarithm on the right side of the equation.	$\ln 2 = 0.1t$
Solve for t and write the exact logarithmic answer.	$t = \dfrac{\ln 2}{0.1}$
Use your calculator to find the decimal answer.	$t = 6.93147$
Round the answer to the nearest year.	$t \approx 7$ years

Radioactive Decay   Formula: $A = A_0 e^{-rt}$

(Observe that the exponent in a decay formula is *negative*.)

EXAMPLE 8   In science, the amount of carbon-14 which remains in a plant or animal after it dies decreases continuously by radioactive decay according to the formula $A = A_0 e^{-0.000124t}$. Use this formula to find the **half-life** of carbon-14 to 3 significant digits. That is, find how long it takes for carbon-14 to decrease to half of its original mass.

Answer:

The initial amount of carbon-14 is not given. Therefore let the original amount $A_0 = 1$ gram. Since the half-life indicates that at the end of t years the carbon-14 present will be half of its original amount, then A = 0.5 gram.

Substitute the numbers into the formula $A = A_0 e^{-0.000124t}$.	$0.5 = 1e^{-0.000124t}$
Simplify the right side of the equation.	$0.5 = e^{-0.000124t}$
The base is *e*. Use *ln* on both sides of the equation.	$\ln 0.5 = \ln e^{-0.000124t}$
Use the Exponent Rule to multiply the exponent times the logarithm.	$\ln 0.5 = (-0.000124t)\ln e$
Cancel out the logarithm on the right side of the equation.	$\ln 0.5 = -0.000124t$
Solve for t and write the exact logarithmic answer.	$t = \dfrac{\ln 0.5}{-0.000124}$
Use your calculator to find the decimal answer.	$t = 5,589.8966$
Round the answer to the nearest year.	$t \approx 5,590$ years

Chapter 8　　Systems of Linear Equations

8.1　THE GRAPHING METHOD

SYSTEMS OF EQUATIONS

A **system** is a set of two or more equations using two or more variables. For a **linear system in two variables x and y**, the equations have no exponents and no variable in the denominator of a fraction. The objective is to find the ordered pair (x, y) which will make both equations true.

The ordered pair (x, y) is the **solution set** for the system.

EXAMPLE 1　　Verify that the ordered pair (3, 2) is the solution set for the linear system:
$$x + y = 5$$
$$x - y = 1$$

Answer:

Using the ordered pair (3, 2), substitute 3 in place of x and 2 in place of y for each equation.

$3 + 2 = 5$　　　(True)

$3 - 2 = 1$　　　(True)

Therefore (3, 2) is the solution set for the system.

SOLVING A LINEAR SYSTEM BY GRAPHING

Given a system of two linear equations in two variables x and y, the solution set may be determined by the **Graphing Method**. We will use the Slope-Intercept Method (from Section 2.4) to *graph both equations on the same rectangular system*. The point where the lines intersect is the solution set.

Solving Systems of Linear Equations – The Graphing Method

For the first equation:

(a) Solve the equation for y.
(b) Write the slope and y-intercept.
(c) Use the slope and y-intercept to neatly draw a straight line graph.
　　　Make sure the graph crosses the x-axis. Label the graph.

For the second equation:

(d) Solve the equation for y.
(e) Write the slope and y-intercept.
(f) Place a point at the y-intercept. Count the slope until you cross the first line.
　　　Neatly draw the second line and label it.

(g) The solution is the point of intersection. Write the solution as an ordered pair.

(8.1)

EXAMPLE 2   Solve by graphing: $3x - 2y = 12$
$7x + 2y = 8$

Answer:

(a) For the first equation, solve for y.

$3x - 2y = 12$

　　Subtract 3x from both sides of the equation.

$-2y = -3x + 12$

　　Divide each term by -2.

$y = \dfrac{3}{2}x - 6$

(b) The slope $m = \dfrac{3}{2} = \dfrac{\text{rise}}{\text{run}}$. The y-intercept is $b = -6$ which is the ordered pair $(0, -6)$.

(c) Neatly draw the graph. Count the slope until the line crosses the x-axis.

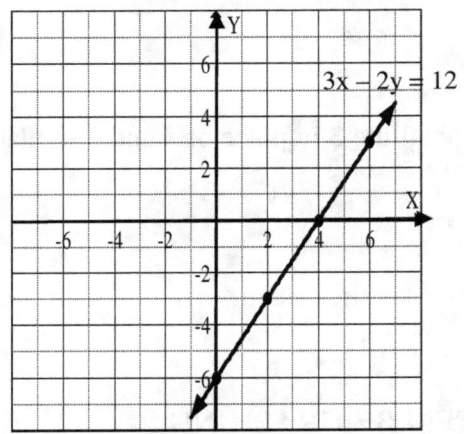

(d) For the second equation, solve for y.

$7x + 2y = 8$

　　Subtract 7x from both sides of the equation.

$2y = -7x + 8$

　　Divide each term by 2.

$y = -\dfrac{7}{2}x + 4$

(e) The slope $m = \dfrac{-7}{2} = \dfrac{\text{rise}}{\text{run}}$. The y-intercept is $b = 4$ which is the ordered pair $(0, 4)$.

(f) Neatly graph the second equation on the same rectangular system; count the slope until the second line crosses the first line.

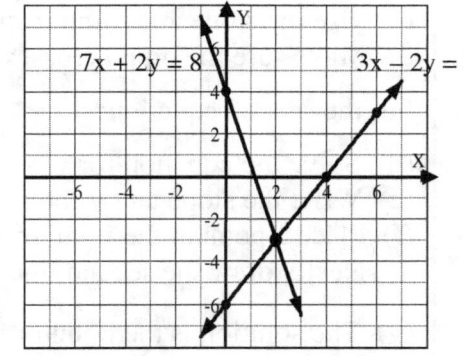

(g) The solution is the point of intersection, $(2, -3)$.

Solution set: $x = 2$, $y = -3$

EXAMPLE 3   Solve by graphing: $y = \dfrac{1}{2}x + 1$

$$x - 2y = -6$$

Answer:

(a) The first equation, $y = \dfrac{1}{2}x + 1$, is already solved for y.

(b) The slope $m = \dfrac{1}{2} = \dfrac{\text{rise}}{\text{run}}$. The y-intercept is $b = 1$ which is the ordered pair $(0, 1)$.

(c) Neatly draw the graph. Count the slope until the line crosses the x-axis.

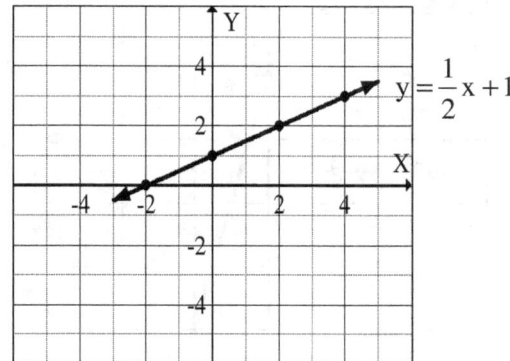

(d) For the second equation, solve for y. $x - 2y = -6$

   Subtract x from both sides of the equation. $-2y = -x - 6$

   Divide each term by -2. $y = \dfrac{x}{2} + 3$

   $$y = \dfrac{1}{2}x + 3$$

(e) The slope $m = \dfrac{1}{2} = \dfrac{\text{rise}}{\text{run}}$. The y-intercept is $b = 3$ which is the ordered pair $(0, 3)$.

(f) Neatly graph the second equation on the same rectangular system.

(g) The lines are parallel; they have the same slope.
   (i.e. The lines do not intersect.)

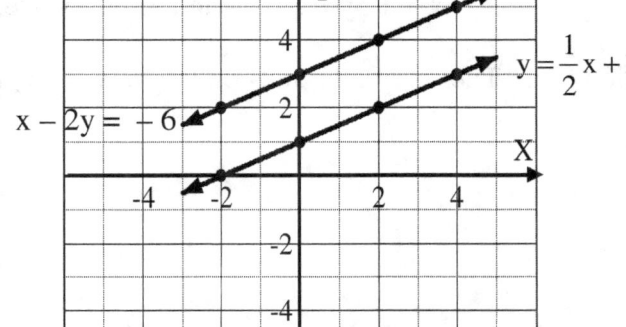

Result: This is an **inconsistent system.**
   This system has *No Solution*.

(8.1)

EXAMPLE 4   Solve by graphing: $y = -4x + 2$
$6x + 1.5y = 3$

Answer:

(a) The first equation, $y = -4x + 2$, is already solved for y.

(b) The slope m = $\dfrac{-4}{1} = \dfrac{rise}{run}$. The y-intercept is b = 2 which is the ordered pair (0, 2).

(c) Neatly draw the graph. Count the slope until the line crosses the x-axis.

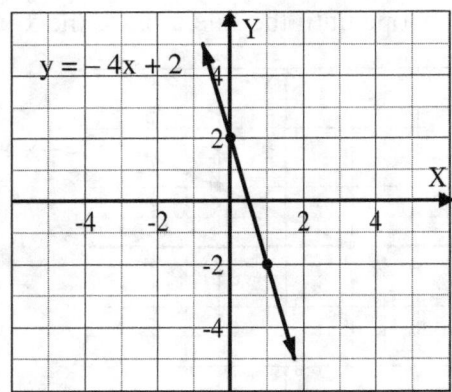

(d) For the second equation, solve for y. $6x + 1.5y = 3$

Subtract 6x from both sides of the equation. $1.5y = -6x + 3$

Divide each term by 1.5. $\dfrac{1.5y}{1.5} = \dfrac{-6x}{1.5} + \dfrac{3}{1.5}$

(e) In slope-intercept form, the second equation is the same as $y = -4x + 2$
the first equation and the graph is the same line.

(f) In other words *the second line coincides with the first line* and each point of the second equation is also a point for the first equation.

The linear system $y = -4x + 2$
$6x + 1.5y = 3$ is a **dependent system**. This system has ***infinite solutions***.

Solution Set: The set of all points on the line, $\{(x,y) \mid y = -4x + 2\}$.

8.2   THE SUBSTITUTION METHOD

An alternative to the Graphing Method is the **Substitution Method**. A system of linear equations in two variables may be solved by the Substitution Method if one of the equations is solved for one variable:

$$x = \sim\ \sim\ \sim \qquad or \qquad y = \sim\ \sim\ \sim\ \sim$$

Solving Systems of Linear Equations – The Substitution Method

- Start with the equation which has one variable on the left side of the equal sign.

- Substitute the expression after the equal sign into the other equation in place of the appropriate variable.

- Solve the resulting equation to find the answer for one of the variables.

- Choose one of the original equations. Substitute the first answer in place of the appropriate variable in the equation.

- Solve the resulting equation to find the answer for the other variable.

- You may write the solution as an ordered pair.

EXAMPLE 1   Solve by substitution: $2x - y = 3$
$x = -2y + 14$

Answer:

The second equation, $x = -2y + 14$, is solved for x.

Substitute the expression $-2y + 14$ into the first equation
in place of x. (Use parentheses around $-2y + 14$). $\qquad\qquad 2(-2y + 14) - y = 3$

Solve this equation for y.

Simplify the parentheses. $\qquad\qquad\qquad\qquad\qquad\qquad -4y + 28 - y = 3$

Add the similar terms. $\qquad\qquad\qquad\qquad\qquad\qquad\qquad -5y + 28 = 3$

Subtract 28 from both sides of the equation. $\qquad\qquad\qquad\qquad -5y = -25$

Divide both sides by -5. $\qquad\qquad\qquad\qquad\qquad\qquad\qquad\qquad y = 5$

Choose one of the original equations. (The second equation is easier to use.) $\qquad x = -2y + 14$

Substitute the first answer 5 in place of y. (Use parentheses). $\qquad\qquad x = -2(5) + 14$

Find the answer for x. $\qquad\qquad\qquad\qquad\qquad\qquad\qquad\qquad x = -10 + 14$

$$x = 4$$

Solution set: $x = 4$, $y = 5$
As an ordered pair, the solution is $(4, 5)$.

(8.2)

✠ *Helpful Hints* – (a) When solving a system of two equations in two variables x and y remember to find the answer for both x and y. (b) In the previous example, if the system had been solved by the Graphing Method, the lines would intersect at the point (4,5). *Observe that the solution for a system of equations is the same regardless of the method used.*

EXAMPLE 2   Solve by the Substitution Method: $y = \dfrac{1}{3}x - 3$
$$x + 3y = 5$$

Answer:

The first equation, $y = \dfrac{1}{3}x - 3$, is solved for y.

Substitute the expression $\dfrac{1}{3}x - 3$ into the second equation $\qquad\qquad x + 3\left(\dfrac{1}{3}x - 3\right) = 5$

in place of y.

Solve this equation for x. $\qquad\qquad\qquad\qquad\qquad\qquad x + \cancel{(3)}\left(\dfrac{1}{\cancel 3}x\right) - (3)(3) = 5$

Simplify the parentheses. $\qquad\qquad\qquad\qquad\qquad\qquad\qquad x + x - 9 = 5$

Add the similar terms. $\qquad\qquad\qquad\qquad\qquad\qquad\qquad\quad 2x - 9 = 5$

Add 9 to both sides of the equation. $\qquad\qquad\qquad\qquad\qquad\qquad 2x = 14$

Divide both sides by 2. $\qquad\qquad\qquad\qquad\qquad\qquad\qquad\qquad x = 7$

Choose one of the original equations. (The second equation is easier to use.) $\qquad x + 3y = 5$

Substitute the first answer 7 in place of x. Solve this equation for y. $\qquad 7 + 3y = 5$

Subtract 7 from both sides of the equation. $\qquad\qquad\qquad\qquad\qquad 3y = -2$

Divide both sides by 3. $\qquad\qquad\qquad\qquad\qquad\qquad\qquad\qquad y = -\dfrac{2}{3}$

Solution set: $x = 7$, $y = -\dfrac{2}{3}$. As an ordered pair, the solution is $\left(7, -\dfrac{2}{3}\right)$.

EXAMPLE 3   Solve by the Substitution Method: $x = 2 - 3y$
$$2x + 6y = 3$$

Answer:

The first equation, $x = 2 - 3y$, is solved for x.

Substitute the expression $2 - 3y$ into the first equation
in place of x. $\qquad\qquad\qquad\qquad\qquad\qquad\qquad\qquad 2(2 - 3y) + 6y = 3$

Solve this equation for y.

Simplify the parentheses. $\qquad\qquad\qquad\qquad\qquad\qquad\qquad 4 - 6y + 6y = 3$

Combine the similar terms. $\qquad\qquad\qquad\qquad\qquad\qquad\qquad\qquad 4 = 3$

The variable terms canceled out and $4 \neq 3$. $\qquad\qquad\qquad$ This is FALSE.

This is an **inconsistent system**. This system has *No Solution*.

(8.2)

✴ ***Helpful Hint*** – When solving a system of equations: If the variable terms cancel out and the numerical result is *false*, then the final answer is *no solution*. (It also indicates that if the Graphing Method had been used, the graphs would not intersect.)

EXAMPLE 4   Solve by the Substitution Method: $3x - 5y = 10$
$$y = 0.6x - 2$$

Answer:

The second equation, $y = 0.6x - 2$, is solved for y.

Substitute the expression $0.6x - 2$ into the first equation
  in place of y. $3x - 5(0.6x - 2) = 10$

Solve this equation for x.

  Simplify the parentheses. $3x - 3x + 10 = 10$

  Combine the similar terms. $10 = 10$

The variable terms canceled out and the result is a true statement.   This is TRUE.

This is a **dependent system**. It has ***infinite solutions***. In slope-intercept form, the first equation is the same as the second equation:

$$3x - 5y = 10 \;\rightarrow\; -5y = -3x + 10 \;\rightarrow\; y = \frac{3}{5}x - 2 \;\rightarrow\; y = 0.6x - 2$$

Solution Set: The set of all points on the line, $\{(x, y) \mid y = 0.6x - 2\}$.

✴ ***Helpful Hint*** – When solving a system of equations: If the variable terms cancel out and the numerical result is *true*, then the system of equations has *infinite solutions*. (It also indicates that if the Graphing Method had been used, the graphs would coincide; the graph of both equations would be the same.)

APPLICATIONS

EXAMPLE 5   Write a linear system of two equations in two variables and solve it by the Substitution Method: The sum of two numbers is 34. Second number is 7 more than twice the first number. Find both numbers.

First number   x

Second number   y

Linear system: $x + y = 34$ "The sum of two numbers is 34."
                      $y = 2x + 7$ "Second number is 7 more than twice the first number."

The second equation, $y = 2x + 7$, is solved for y.

Substitute the expression $2x + 7$ into the first equation
  in place of y. $x + (2x + 7) = 34$

(8.2)

Solve this equation for *x*.	$x + 2x + 7 = 34$
Add the similar terms.	$3x + 7 = 34$
Subtract 7 from both sides of the equation.	$3x = 27$
Divide both sides by 3.	$x = 9$
Choose one of the original equations. (The first equation is easier to use.)	$x + y = 34$
Substitute the first answer 9 in place of *x*.	$9 + y = 34$
Solve this equation for *y*.	$y = 25$

Solution set: $x = 9, y = 25$

EXAMPLE 6   Write a linear system of two equations in two variables and solve it by the Substitution Method:
In a rectangle, the length is 5 feet less than 4 times the width. The perimeter is 110 centimeters. Find the length and width.

length   L

width   W

Perimeter   110 cm

Perimeter formula $P = 2L + 2W$

Linear system: $L = 4W - 5$ "The length is 5 feet less than 4 times the width."
                  $110 = 2L + 2W$ "The perimeter is 110 centimeters."

The first equation, $L = 4W - 5$, is solved for *L*.

Substitute the expression $4W - 5$ into the second equation in place of *L*. Solve this equation for *W*.	$110 = 2(4W - 5) + 2W$
Simplify the parentheses.	$110 = 8W - 10 + 2W$
Add the similar terms.	$110 = 10W - 10$
Add 10 from both sides of the equation.	$120 = 10W$
Divide both sides by 10.	$W = 12$ cm
Choose one of the original equations. (The first equation is easier to use.)	$L = 4W - 5$
Substitute the first answer 12 in place of *W*.	$L = 4(12) - 5$
Find the answer for *L*.	$L = 43$ cm

Answers: The length of the rectangle is 43 cm and the width is 12 cm.

8.3   THE ADDITION METHOD

Another alternative to the Graphing Method is the the **Addition Method**. A system of linear equations in two variables may be solved by the Addition Method if the equations are written in a "column" format, such as:

$$x - y = -1$$
$$-x + 3y = 7$$

✠ *Helpful Hint* – For the Addition Method, one of the variables must cancel out by adding the equations. Each equation must have the same variable term with different signs. In the above system, the x-column will cancel out when the equations are added.

Solving Systems of Linear Equations – The Addition Method

- Cancel out one of the variables by adding both equations together.

- Solve the resulting equation to find one of the answers.

- Choose one of the original equations. Substitute the first answer in place of the appropriate variable in the equation.

- Solve the resulting equation to find the answer for the other variable.

- You may write the solution as an ordered pair.

EXAMPLE 1   Solve by the Addition Method: $x - y = -1$
$-x + 3y = 7$

Answer:

Cancel out x by adding both equations together.	$2y = 6$
Solve the resulting equation for y.	$y = 3$
Choose one of the original equations. (The first equation is easier to use.)	$x - y = -1$
Substitute the first answer 3 in place of y.	$x - 3 = -1$
Solve this equation for x.	$x = 2$

Solution set: $x = 2, y = 3$

Remember: If the system had been solved by the Graphing Method, the lines would intersect at the point (2, 3). Also, the solution for a system of equations is the same regardless of the method used.

✠ *Helpful Hints* – (1) For the Addition Method, choose a variable to cancel out; you may select either variable and get the same the solution set. (2) If the selected variable does not cancel out by adding the equations, you may multiply one of the equations times a real number which will cause one variable to cancel out. (3) Read Example 2 and Example 3 which show how to use a multiplier to solve a system of equations.

(8.3)

EXAMPLE 2   Using the Addition Method, cancel out the *x-column* and solve:

$$x + 2y = -5$$
$$3x - y = -8$$

Answer:

If we add the equations together, the x-column will *not* cancel out. By multiplying the first equation times -3, then $-3x$ and $3x$ will cancel out.

Multiply the first equation times -3.

$$-3(x + 2y = -5)$$
$$3x - y = -8$$

Rewrite the system.

$$-3x - 6y = 15$$
$$\underline{3x - y = -8}$$

Cancel out x by adding both equations together. $-7y = 7$

Solve the resulting equation for *y*. $y = -1$

Choose one of the original equations. $x + 2y = -5$

Substitute the first answer -1 in place of *y*. $x + 2(-1) = -5$

Solve this equation for *x*. $x - 2 = -5$

    $x = -3$

Solution set: $x = -3, y = -1$. As an ordered pair, the solution is $(-3, -1)$.

EXAMPLE 3   Using the Addition Method, cancel out the *y-column* and solve:

$$x + 2y = -5$$
$$3x - y = -8$$

Answer:

If we add the equations together, the y-column will *not* cancel out. By multiplying the second equation times 2, then $2y$ and $-2y$ will cancel out.

Multiply the second equation times 2.

$$x + 2y = -5$$
$$2(3x - y = -8)$$

Rewrite the system.

$$x + 2y = -5$$
$$\underline{6x - 2y = -16}$$

Cancel out *y* by adding both equations together. $7x = -21$

Solve the resulting equation for *x*. $x = -3$

Choose one of the original equations. $x + 2y = -5$

Substitute the first answer -3 in place of *x*. $-3 + 2y = -5$

Solve this equation for *y*. $2y = -2$

    $y = -1$

Solution set: $x = -3, y = -1$.

Remember, you may cancel either variable and the resulting solution set will be the same.

✖ **_Helpful Hints_** – (1) To cancel out the selected variable, you may multiply the first equation times a number and multiply the second equation times a different number which will cause the selected variable to cancel out when the equations are added together. (2) It is usually easier to choose the variable which has smaller coefficients.

EXAMPLE 4   Solve by the Addition Method: $11x + 2y = 1$
$$5x - 3y = 20$$
Answer:

If we add the equations together, neither the x-column nor the y-column will *not* cancel out. To cancel out the x-column, we could multiply the first equation times 5 and multiply –11 times the second equation. Then 55x and – 55x would cancel out. However, the coefficients of *y* are smaller numbers. Therefore we choose to cancel out the y-column.

Multiply the first equation times 3 and multiply the second equation times 2.	$3(11x + 2y = 1)$ $2(5x - 3y = 20)$
Rewrite the system.	$33x + 6y = 3$ $\underline{10x - 6y = \;\;40}$
Cancel out *y* by adding both equations together.	$43x = 43$
Solve the resulting equation for *x*.	$x = 1$
Choose one of the original equations.	$11x + 2y = 1$
Substitute the first answer 1 in place of *x*.	$11(1) + 2y = 1$
Solve this equation for *y*.	$11 + 2y = 1$
	$2y = -10$
	$y = -5$

Solution set: $x = 1, y = -5$.

EXAMPLE 5   Solve by the Addition Method: $x - 3y = 2$
$$2x - 6y = 7$$
Answer:

Choose one of the variables, such as *x*. (The x-column is easier.) Multiply the first equation times – 2.	$-2(x - 3y = 2)$ $2x - 6y = 7$
Rewrite the system.	$-2x + 6y = -4$ $\underline{2x - 6y = \;\;7}$
Add both equations together.	$0 = 3$
Both variables canceled out and $0 \neq 3$.	This is FALSE.

This is an **inconsistent system**. This system has *No Solution*.

(8.3)

EXAMPLE 6   Solve by the Addition Method: $\frac{1}{4}x + y = -2$
$$-x - 4y = 8$$

Answer:

Choose one of the variables, such as y. (The y-column is easier.)

Multiply the first equation times 4.

$$4(\frac{1}{4}x + y = -2)$$
$$-x - 4y = 8$$

Rewrite the system.

$$x + 4y = -8$$
$$\underline{-x - 4y = 8}$$

Adding both equations together.

$$0 = 0$$

Both variables canceled out and the result is a true statement.   This is TRUE.

This is a **dependent system**. It has *infinite solutions*. In slope-intercept form, the first equation is the same as the second equation:

$$\frac{1}{4}x + y = -2 \;\rightarrow\; y = -\frac{1}{4}x - 2$$

$$-x - 4y = 8 \;\rightarrow\; -4y = x + 8 \;\rightarrow\; y = -\frac{1}{4}x - 2$$

Solution Set: The set of all points on the line, $\{(x, y) \mid y = -\frac{1}{4}x - 2\}$.

APPLICATIONS

EXAMPLE 7   Write a linear system of two equations in two variables and solve it by the Addition Method: The sum of two numbers is 17. Their difference is 89. Find both numbers.

First number   x

Second number   y

Linear system: $x + y = 17$ "The sum of two numbers is 17."
$x - y = 89$ "Their difference is 89."

Cancel out y by adding both equations together. $2x = 106$

Solve the resulting equation for x. $x = 53$

Choose one of the original equations. (The first equation is easier to use.) $x + y = 17$

Substitute the first answer 53 in place of x. $53 + y = 17$

Solve this equation for y. $y = -36$

Solution set: $x = 53, y = -36$

(8.3)

EXAMPLE 8   Write a linear system of two equations in two variables and solve it by the Addition Method:

Sarah sold 27 tickets to the Community Center Drama Society's summer play. The tickets are $10 for adults and $5 for children. Sarah turned in a total of $230. How many of each kind of ticket did she sell?

Number of adult tickets   x

Number of children's tickets   y

Total number of tickets   27

Dollar value of the adult tickets   10x   (The tickets are $10 each for adults.)

Dollar value of the children's tickets   5y   (The tickets are $5 each for children.)

Total amount for the tickets sold $230

Linear System: $x + y = 27$ "Sarah sold 27 tickets."

$10x + 5y = 230$ "Sarah turned in a total of $230."

Choose one of the variables, such as x.
Multiply the first equation times -10.

$$-10(x + y = 27)$$
$$10x + 5y = 230$$

Rewrite the system.

$$-10x - 10y = -270$$
$$\underline{10x + 5y = 230}$$

Cancel out x by adding both equations together. $-5y = -40$

Solve the resulting equation for y. $y = 8$

Choose one of the original equations. (The first equation is easier to use.) $x + y = 27$

Substitute the first answer 53 in place of y. $x + 8 = 27$

Solve this equation for x. $x = 19$

Solution set: $x = 19$ adult tickets, $y = 8$ children's tickets

Check:

Dollar value of the adult tickets $10(19) = \$190$

Dollar value of the children's tickets $5(8) = \$40$

Total amount for the tickets sold $190 + 40 = \$230$

8.4   SYSTEMS OF LINEAR EQUATIONS IN THREE VARIABLES

LINEAR EQUATIONS IN THREE VARIABLES

The equation $x + 2y - z = 7$ is called a **linear equation in three variables**. Each solution to an equation in three variables is an **ordered triple (x, y, z)** which will make the equation true.

EXAMPLE 1   Determine whether each ordered triple is a solution for the linear equation
$x + 2y - z = 7$: (a) (2, 3, 1)   (b) $(-5, 6, 0)$ (c) $(1, -4, -2)$

Answers:

(a) For (2, 3, 1), substitute 2 in place of x, 3 in place of y and 1 in place of z in the equation.
$2 + 2(3) - 1 = 7$
$2 + 6 - 1 = 7$ is true. Therefore (2, 3, 1) is a solution for the equation $x + 2y - z = 7$.

(b) For $(-5, 6, 0)$, substitute -5 in place of x, 6 in place of y and 0 in place of z in the equation.
$-5 + 2(6) - 0 = 7$
$-5 + 12 - 0 = 7$ is true. Therefore $(-5, 6, 0)$ is a solution for $x + 2y - z = 7$.

(a) For $(1, -4, -2)$, substitute 1 in place of x, -4 in place of y and -2 in place of z in the equation.
$1 + 2(-4) - (-2) = 7$
$1 - 8 + 2 = 7$
$-5 = 7$ is *false*. Therefore $(1, -4, -2)$ is *not* a solution for $x + 2y - z = 7$.

SYSTEMS OF LINEAR EQUATIONS IN THREE VARIABLES

Given a linear system of three equations in three variables x, y and z, the objective is to find the ordered triple (x, y, z) which will make all of the equations true. The ordered triple (x, y, z) is the solution set for the system.

EXAMPLE 2   Verify that the ordered pair $(3, -2, 1)$ is the solution set for the linear system:
$$x + y + z = 2$$
$$x - 2y + z = 8$$
$$x + 3y - 4z = -7$$

Answer:

For $(3, -2, 1)$, substitute 3 in place of x, -2 in place of y and 1 in place of z for each equation.

$3 + (-2) + 1 = 2$	$3 - 2(-2) + 1 = 8$	$3 + 3(-2) - 4(1) = -7$
$3 - 2 + 1 = 2$	$3 + 4 + 1 = 8$	$3 - 6 - 4 = -7$
(True)	(True)	(True)

Therefore $(3, -2, 1)$ is the solution set for the system.

SOLVING SYSTEMS OF LINEAR EQUATIONS IN THREE VARIABLES

A linear system of three equations in three variables may have one solution, (x, y, z), or no solution, or infinite solutions. To solve the system of equations, we will use either Substitution or the Addition Method, similar to the procedures shown in sections 8.2 and 8.3.

(8.4)

Using Substitution To Solve A Linear System in Three Variables

If one of the equations is solved for one variable, use the Substitution Method to solve the linear system.

EXAMPLE 3   Solve the linear system:
$$x + y + z = 7$$
$$2y + z = 11$$
$$z = 5$$

Answer:

The third equation is already solved for z. We will use Substitution to solve the system.

In the second equation, substitute 5 in place of z. $2y + 5 = 11$

Solve this equation for y. $2y = 6$

                                                                                                                                    $y = 3$

In the first equation, substitute 3 in place of y and 5 in place of z. $x + 3 + 5 = 7$

Solve this equation for x. $x + 8 = 7$

                                                                                                                                   $x = -1$

The solution set is the ordered triple $(-1, 3, 5)$.

✖ *__Helpful Hint__* – Write an ordered triple with the x-value first, the y-value second and the z-value third. Also observe that an ordered triple must be in parentheses.

Using The Addition Method To Solve A Linear System in Three Variables

If the equations are written in a "column" format, use the Addition Method to solve the linear system.

Solving 3 Linear Equations in 3 Variables – The Addition Method

- For the first two equations, use the Addition procedure cancel out one of the variables. The result is equation [E4].
- For the second and third equations, use the Addition procedure cancel out the same variable. The result is equation [E5].

- Solve the resulting linear system, equations [E4] and [E5]. Find the answer for two variables.

- Choose one of the original equations. Substitute the both answers in place of the appropriate variables in the equation.

- Solve the resulting equation to find the answer for the remaining variable.

- Write the solution as an ordered triple (x, y, z).

(8.4)

EXAMPLE 4   Solve the linear system:

$$2x + 2y - 3z = -2$$
$$x - y - z = 5$$
$$x - 3y + 4z = 11$$

Answer:

Number the equations.

[E1] $2x + 2y - 3z = -2$
[E2] $x - y - z = 5$
[E3] $x - 3y + 4z = 11$

For equations [E1] and [E2], use the Addition procedure to cancel out the x-column. The result is equation [E4].

$$2x + 2y - 3z = -2$$
$$-2(x - y - z = 5)$$

$$2x + 2y - 3z = -2$$
$$\underline{-2x + 2y + 2z = -10}$$
[E4] $4y - z = -12$

For equations [E2] and [E3], use the Addition procedure to cancel out the x-column. The result is equation [E5].

$$x - y - z = 5$$
$$-1(x - 3y + 4z = 11)$$

$$x - y - z = 5$$
$$\underline{-x + 3y - 4z = -11}$$
[E5] $2y - 5z = -6$

Solve the resulting linear system, equations [E4] and [E5].

[E4] $4y - z = -12$
[E5] $2y - 5z = -6$

To cancel out y, multiply equation [E5] times -2.

$$4y - z = -12$$
$$-2(2y - 5z = -6)$$

$$4y - z = -12$$
$$\underline{-4y + 10z = 12}$$

Find the answer for z.

$$9z = 0$$
$$z = 0$$

Use equation [E4] to find the answer for y.

[E4] $4y - z = -12$

$$4y - 0 = -12$$

$$4y = -12$$

$$y = -3$$

Choose one of the original equations.

[E2] $x - y - z = 5$

Substitute the both answers in place of the appropriate variables in the equation.

$$x - (-3) - 0 = 5$$
$$x + 3 = 5$$

Find the answer for x.

$$x = 2$$

The solution set: The ordered triple $(2, -3, 0)$.

(8.4)

Using Substitution and Addition To Solve A Linear System

EXAMPLE 5   Solve the linear system:

$$x - 3y \qquad = \ 3$$
$$3y + 2z = \ 2$$
$$3x \qquad - 4z = 10$$

Answer:

Number the equations.

[E1] $x - 3y \qquad = \ 3$
[E2] $\qquad 3y + 2z = \ 2$
[E3] $3x \qquad - 4z = 10$

Each equation is missing a different variable. Rewrite equation [E1] and use Substitution.

Solve equation [E1] for x.

$x = 3y + 3$

Using equation [E3], substitute for x and rewrite the equation.

[E3] $3x \quad - 4z = 10$
$3(3y + 3) - 4z = 10$
$9y + 9 - 4z = 10$

The result is equation [E4].

[E4] $9y - 4z = \ 1$

Write the linear system consisting of equations [E2] and [E4].
Use the Addition Method to solve this linear system.

[E2] $3y + 2z = \ 2$
[E4] $9y - 4z = \ 1$

To cancel out z, multiply equation [E2] times 2.

$2(3y + 2z = \ 2)$
$9y - 4z = \ 1$

$6y + 4z = \ 4$
$\underline{9y - 4z = \ 1}$

Find the answer for y.

$15y \qquad = \ 5$

$y = \dfrac{1}{3}$

Use equation [E2] to find the answer for z.

[E2] $3y + 2z = \ 2$

$\cancel{3}\left(\dfrac{1}{\cancel{3}}\right) + 2z = \ 2$

$1 + 2z = \ 2$

$2z = \ 1$

$z = \dfrac{1}{2}$

Choose one of the original equations.

[E1] $x - 3y = \ 3$

Substitute the answer for y in the equation.

$x - 3\left(\dfrac{1}{3}\right) = \ 3$

$x - 1 = 3$

Find the answer for x.

$x = 4$

The solution set: The ordered triple $\left(4, \dfrac{1}{3}, \dfrac{1}{2}\right)$.

(8.4)

Inconsistent and Dependent Systems of Linear Equations in Three Variables

EXAMPLE 6   Solve the linear system:

$$x - y - z = -3$$
$$x - 5y + z = 6$$
$$x + 5y - 4z = 5$$

Answer:

Number the equations.

[E1] $x - y - z = -3$
[E2] $x - 5y + z = 6$
[E3] $x + 5y - 4z = 5$

For equations [E1] and [E2], use the Addition procedure to cancel out the x-column. The result is equation [E4].

$$-1(x - y - z = -3)$$
$$x - 5y + z = 6$$

$$-x + y + z = 3$$
$$\underline{x - 5y + z = 6}$$
[E4] $-4y + 2z = 9$

For equations [E1] and [E3], use the Addition procedure to cancel out the x-column. The result is equation [E5].

$$-1(x - y - z = -3)$$
$$x + 5y - 4z = 5$$

$$-x + y + z = 3$$
$$\underline{x + 5y - 4z = 5}$$
[E5] $6y - 3z = 8$

Solve the resulting linear system, equations [E4] and [E5].

[E4] $-4y + 2z = 9$
[E5] $6y - 3z = 8$

To cancel out *y*, multiply equation [E4] times 3 and multiply equation [E4] times 2.

$$3(-4y + 2z = 9)$$
$$2(6y - 3z = 8)$$

$$-12y + 6z = 27$$
$$\underline{12y - 6z = 16}$$
$$0 = 43$$

Add both equations together.

Both variables canceled out and $0 \neq 43$.

This is FALSE.

This is an **inconsistent system**. This system has *No Solution*.

EXAMPLE 7   Solve the linear system:

$$x + y - z = 3$$
$$x - y + z = 1$$
$$x + 3y - 3z = 5$$

Answer:

This system of equations is arranged in a column format. We will use the Addition Method to find the solution set.

(8.4)

Number the equations.

$$[E1] \quad x + y - z = 3$$
$$[E2] \quad x - y + z = 1$$
$$[E3] \quad x + 3y - 3z = 5$$

For equations [E1] and [E2], use the Addition procedure to cancel out the x-column. The result is equation [E4].

$$x + y - z = 3$$
$$-1(x - y + z = 1)$$

$$x + y - z = 3$$
$$\underline{-x + y - z = -1)}$$
$$[E4] \qquad 2y - 2z = 2$$

For equations [E1] and [E3], use the Addition procedure to cancel out the x-column. The result is equation [E5].

$$x + y - z = 3$$
$$-1(x + 3y - 3z = 5)$$

$$x + y - z = 3$$
$$\underline{-x - 3y + 3z = -5}$$
$$[E5] \qquad -2y + 2z = -2$$

Solve the resulting linear system, equations [E4] and [E5].

$$[E4] \qquad 2y - 2z = 2$$
$$[E5] \qquad \underline{-2y + 2z = -2}$$

Add both equations together.

$$0 = 0$$

Both variables canceled out and the result is a true statement.

This is TRUE.

This is a **dependent system**. It has *infinite solutions*.

Let z be any real number.

z = any real number

Solve equation [E4] for y.

$$[E4] \qquad 2y - 2z = 2$$
$$2y = 2z + 2$$
$$y = z + 1$$

Using equation [E1], substitute z + 1 for y.

$$[E1] \qquad x + y - z = 3$$
$$x + (z + 1) - z = 3$$

Solve the equation for x.

$$x + 1 = 3$$
$$x = 2$$

Solution Set: The set of all ordered triples $\{(x, y, z) \mid (x, y, z) = (2, z + 1, z)\}$ where z is any real number.

�֍ *__Helpful Hint__* – The following are some of the ordered triples in the solution set for Example 7.

Value of z	(x, y, z) = (2, z + 1, z)
⋮	⋮
z = −1	(2, 0, −1)
z = 0	(2, 1, 0)
z = 1	(2, 2, 1)
⋮	⋮

Each ordered triple in the solution set makes each equation of the linear system true.

(You should verify this in your notebook.)

8.5   THE GAUSS-JORDAN MATRIX METHOD

MATRIX DEFINITIONS

A **matrix** is a table of numbers enclosed in brackets. The rows of a matrix are across, and the columns are vertical.   The **dimension** of a matrix is the size of a matrix: An **n x p** matrix has **n** rows across and **p** columns vertically. (We say "The dimension of the matrix is n by p.")

EXAMPLE 1   State the dimension of each matrix.

(a) $\begin{bmatrix} 14 & 3 & 6 \\ 12 & 0 & -4 \end{bmatrix}$ is a 2 x 3 matrix.

(b) $\begin{bmatrix} -1 & 4 \\ -8 & 9 \\ 0 & 3 \end{bmatrix}$ is a 3 x 2 matrix.

A **square matrix** is an *n x n* matrix. It has the same number of rows and columns. The following are square matrices:

$\begin{bmatrix} 5 & 4 \\ 3 & -7 \end{bmatrix}$
2 x 2 matrix

$\begin{bmatrix} 10 & 2 & -1 \\ -6 & 13 & 11 \\ 7 & -9 & -21 \end{bmatrix}$
3 x 3 matrix

An **identity matrix**, denoted as I_n , is an *n x n* (square) matrix with 1's on the main diagonal and zeros elsewhere.

$I_2 = \begin{bmatrix} 1 & 0 \\ 0 & 1 \end{bmatrix}$

$I_3 = \begin{bmatrix} 1 & 0 & 0 \\ 0 & 1 & 0 \\ 0 & 0 & 1 \end{bmatrix}$

THE AUGMENTED MATRIX AND ROW OPERATIONS

The Augmented Matrix

Given the system of linear equations, we can set up a special matrix called an **augmented matrix**.
Let $Ax + By = C$ be a linear system in two variables x and y.
$Dx + Ey = F$

The augmented matrix is $\begin{bmatrix} A & B & | & C \\ D & E & | & F \end{bmatrix}$.

The first column has the coefficients of x, the second column has the coefficients of y. After the vertical line, the last column has the constants which were after the equal signs.

Similarly, the linear system $Ax + By + Cz = D$ has the augmented matrix:
$Ex + Fy + Gz = H$
$Jx + Ky + Lz = M$

$\begin{bmatrix} A & B & C & | & D \\ E & F & G & | & H \\ J & K & L & | & M \end{bmatrix}$

EXAMPLE 2   Write the augmented matrix for each system of equations.

(a) $11x + 2y = 1$
$5x - 3y = 20$

(b) $x + y + z = 2$
$x - 2y + z = 8$
$x + 3y - 4z = -7$

Answers

(a)   For the system $11x + 2y = 1$ the augmented matrix is: $\begin{bmatrix} 11 & 2 & | & 1 \\ 5 & -3 & | & 20 \end{bmatrix}$
$\qquad\qquad\qquad\quad\; 5x - 3y = 20$

(b)   For the system $x + y + z = 2$ the augmented matrix is: $\begin{bmatrix} 1 & 1 & 1 & | & 2 \\ 1 & -2 & 1 & | & 8 \\ 1 & 3 & -4 & | & -7 \end{bmatrix}$
$\qquad\qquad\qquad\quad\; x - 2y + z = 8$
$\qquad\qquad\qquad\quad\; x + 3y - 4z = -7$

EXAMPLE 3   Write the system of equations represented by each augmented matrix.

(a) $\begin{bmatrix} 7 & 1 & | & 1 \\ 5 & -2 & | & 17 \end{bmatrix}$ (b) $\begin{bmatrix} 1 & 0 & | & 3 \\ 0 & 1 & | & -2 \end{bmatrix}$ (c) $\begin{bmatrix} 2 & 2 & -3 & | & -2 \\ 1 & -1 & -1 & | & 5 \\ -2 & 0 & 7 & | & 4 \end{bmatrix}$

Answers

For augmented matrix (a), the system of equations is: $7x + y = 1$
$\qquad\qquad\qquad\qquad\qquad\qquad\qquad\qquad\qquad\quad 5x - 2y = 17$

For augmented matrix (b), the system of equations is: $x + 0 = 3 \quad\rightarrow\quad x = 3$
$\qquad\qquad\qquad\qquad\qquad\qquad\qquad\qquad\qquad\quad 0 + y = -2 \quad\rightarrow\quad y = -2$

For augmented matrix (c), the system of equations is: $2x + 2y - 3z = -2$
$\qquad\qquad\qquad\qquad\qquad\qquad\qquad\qquad\qquad\qquad x - y - z = 5$
$\qquad\qquad\qquad\qquad\qquad\qquad\qquad\qquad\quad -2x \qquad + 7z = 4$

Row Operations for the Augmented Matrix

When we solved a system of equations using the Addition Method, we used multiplication and addition to rewrite the system and solve for one of the variables. In this section, we will use similar **row operations** to rewrite the augmented matrix and solve the system of linear equations.

Given an augmented matrix, let R_n and R_p represent two rows of the augmented matrix. Let C be a constant, a real number other than zero.

The following are the **Rules for Row Operations**.

1. $R_n \longleftrightarrow R_p$ Two rows of the matrix may be interchanged (switched).

2. $C \cdot R_n$ A row may be multiplied or divided by number other than zero.
   $\dfrac{R_n}{C}$

3. $R_n + R_p$ A row may be changed by adding it to (or subtract it from) another row.
   $R_n - R_p$

4. $C \cdot R_n + R_p$ A row may be multiplied by a constant and then added to another row.

177

(8.5)

✠ *Helpful Hint* – The row operation "$R_1 + R_2 \rightarrow R_2$" means add row 1 to row 2 and write the result in row 2. Leave row 1 unchanged. Refer to the next example.

EXAMPLE 4   Perform the row operation and rewrite the augmented matrix $\begin{bmatrix} 4 & 1 & | & -2 \\ -1 & 3 & | & 0 \end{bmatrix}$.

(a) $R_1 \longleftrightarrow R_2$ (b) $\dfrac{R_1}{2} \rightarrow R_1$ (c) $-3R_2 + R_1 \rightarrow R_1$

Answers:

(a) $R_1 \longleftrightarrow R_2$ means interchange row 1 and row 2. $\begin{bmatrix} 4 & 1 & | & -2 \\ -1 & 3 & | & 0 \end{bmatrix} \rightarrow \begin{bmatrix} -1 & 3 & | & 0 \\ 4 & 1 & | & -2 \end{bmatrix}$

(b) $\dfrac{R_1}{2} \rightarrow R_1$ means divide row 1 by 2 and write the result in row 1. Row 2 is unchanged.

$$\begin{bmatrix} 4 & 1 & | & -2 \\ -1 & 3 & | & 0 \end{bmatrix} \rightarrow \begin{bmatrix} 2 & \dfrac{1}{2} & | & -1 \\ -1 & 3 & | & 0 \end{bmatrix}$$

(c) $-3R_1 + R_2 \rightarrow R_2$ means for each number on row 1, multiply -3 times the number and add the corresponding number of row 2; write the result on row 2. Row 1 is unchanged.

$$-3R_1 + R_2 \rightarrow R_2 \quad \begin{bmatrix} 4 & 1 & | & -2 \\ -1 & 3 & | & 0 \end{bmatrix} \rightarrow \begin{bmatrix} 4 & 1 & | & -2 \\ -13 & 0 & | & 6 \end{bmatrix}$$

Workspace for the new row 2:
$\begin{array}{ccc} -3(4) & -3(1) & -3(-2) \\ +(-1) = -13 & +3 = 0 & +0 = 6 \end{array}$

THE GAUSS-JORDAN MATRIX METHOD

Solving Systems of Two Linear Equations in Two Variables

Given a system of linear equations $\begin{aligned} Ax + By &= C \\ Dx + Ey &= F, \end{aligned}$

$\downarrow$

set up the augmented matrix: $\begin{bmatrix} A & B & | & C \\ D & E & | & F \end{bmatrix}$

Use the row operations to get $\downarrow$
     the **solution matrix:** $\downarrow$

$$\begin{bmatrix} 1 & 0 & | & m \\ 0 & 1 & | & p \end{bmatrix}$$

Therefore the solution set is: $x = \mathbf{m}$ and $y = \mathbf{p}$.

✠ *Helpful Hints* – (1) Observe that the row operations are used to transform the augmented matrix into the solution matrix $\begin{bmatrix} 1 & 0 & | & m \\ 0 & 1 & | & p \end{bmatrix}$ which has the I_2 identity matrix $\begin{bmatrix} 1 & 0 \\ 0 & 1 \end{bmatrix}$ to the left of the vertical line.

(2) An easy way to obtain the solution matrix is to modify the augmented matrix in this order: Get 1 in the first position of row 1, get 0 in the first position of row 2, get 1 in the second position of row 2, and get 0 in the second position of row 1. (3) Then use the solution matrix to write the solution set: x = m and y = p. Refer to the next example.

EXAMPLE 5   Solve by using the Gauss–Jordan Method: x + y = 5
                                                          x – y = 1

Answer:

(a) Set up the augmented matrix: $\begin{bmatrix} 1 & 1 & | & 5 \\ 1 & -1 & | & 1 \end{bmatrix}$

(b) The first position of row 1 is already **1**. Use matrix (a) to get **0** in the first position of row 2. Multiply –1 times row 1 and add to row 2; write the result on row 2:

$$-1 \cdot R_1 + R_2 \to R_2 \quad \begin{bmatrix} 1 & 1 & | & 5 \\ 1 & -1 & | & 1 \end{bmatrix} \to \begin{bmatrix} 1 & 1 & | & 5 \\ \mathbf{0} & \mathbf{-2} & | & \mathbf{-4} \end{bmatrix}$$

$$\begin{array}{ccc} -1(1) & -1(1) & -1(5) \\ +1 = \mathbf{0} & +(-1) = -2 & +1 = -4 \end{array}$$

 (c) Use matrix (b) to get **1** in the second position of row 2. Divide the second row by – 2 and write the result on row 2:

$$\frac{R_2}{-2} \to R_2 \quad \begin{bmatrix} 1 & 1 & | & 5 \\ 0 & -2 & | & -4 \end{bmatrix} \to \begin{bmatrix} 1 & 1 & | & 5 \\ \mathbf{0} & \mathbf{1} & | & \mathbf{2} \end{bmatrix}$$

$$\frac{0}{-2} = 0 \qquad \frac{-2}{-2} = 1 \qquad \frac{-4}{-2} = 2$$

(d) Use matrix (c) to get **0** in the second position of row 1. Multiply –1 times row 2 and add to row 1; write the result on row 1:

$$-1 \cdot R_2 + R_1 \to R_1 \quad \begin{bmatrix} 1 & 1 & | & 5 \\ 0 & 1 & | & 2 \end{bmatrix} \to \begin{bmatrix} \mathbf{1} & \mathbf{0} & | & \mathbf{3} \\ 0 & 1 & | & 2 \end{bmatrix}$$

$$\begin{array}{ccc} -1(0) + 1 = 1 & -1(1) + 1 = \mathbf{0} & -1(2) + 5 = 3 \end{array} \qquad \text{This is the solution matrix.}$$

(e) The solution set: x = 3, y = 2. You may also write the solution set as the ordered pair (3,2).

✠ *Helpful Hints* – (1) When solving a linear system by the Gauss-Jordan Method, you decide the row operations to be used which will help you to obtain the solution matrix. (2) In the previous example, if the system had been solved by graphing, the lines would intersect at the point (3,2). Remember that the solution for a system of equations is the same regardless of the method used.

(8.5)

EXAMPLE 6   Solve by using the Gauss–Jordan Method: $3x - 5y = -20$
$$2x + 11y = 5$$

Answer:

(a) Set up the augmented matrix:
$$\begin{bmatrix} 3 & -5 & -20 \\ 2 & 11 & 1 \end{bmatrix}$$

(b) Use matrix (a) to get **1** in the first position of row 1. Subtract row 2 from row 1; write the result on row 1:

$$R_1 - R_2 \rightarrow R_1 \qquad \begin{bmatrix} 3 & -5 & \bigm| & -20 \\ 2 & 11 & \bigm| & 1 \end{bmatrix} \quad \rightarrow \quad \begin{bmatrix} \mathbf{1} & \mathbf{-16} & \bigm| & \mathbf{-21} \\ 2 & 11 & \bigm| & 1 \end{bmatrix}$$

$$\begin{array}{ccc} 3 & -5 & -20 \\ -2 = \mathbf{1} & -11 = -16 & -1 = -21 \end{array}$$

(c) Use matrix (b) to get **0** in the first position of row 2. Multiply –2 times row 1 and add to row 2; write the result on row 2:

$$-2R_1 + R_2 \rightarrow R_2 \qquad \begin{bmatrix} 1 & -16 & \bigm| & -21 \\ 2 & 11 & \bigm| & 1 \end{bmatrix} \quad \rightarrow \quad \begin{bmatrix} 1 & -16 & \bigm| & -21 \\ \mathbf{0} & \mathbf{43} & \bigm| & \mathbf{43} \end{bmatrix}$$

$$\begin{array}{ccc} -2(1) & -2(-16) & -2(-21) \\ +2 = \mathbf{0} & +11 = 43 & +1 = 43 \end{array}$$

(d) Use matrix (c) to get **1** in the second position of row 2. Divide the second row by 43 and write the result on row 2:

$$\frac{R_2}{43} \rightarrow R_2 \qquad \begin{bmatrix} 1 & -16 & \bigm| & -21 \\ 0 & 43 & \bigm| & 43 \end{bmatrix} \quad \rightarrow \quad \begin{bmatrix} 1 & -16 & \bigm| & -21 \\ \mathbf{0} & \mathbf{1} & \bigm| & \mathbf{1} \end{bmatrix}$$

$$\frac{0}{43} = 0 \qquad \frac{43}{43} = \mathbf{1} \qquad \frac{43}{43} = 1$$

(e) Use matrix (d) to get **0** in the second position of row 1. Multiply 16 times row 2 and add to row 1; write the result on row 1:

$$16R_2 + R_1 \rightarrow R_1 \qquad \begin{bmatrix} 1 & -16 & \bigm| & -21 \\ 0 & 1 & \bigm| & 1 \end{bmatrix} \quad \rightarrow \quad \begin{bmatrix} \mathbf{1} & \mathbf{0} & \bigm| & \mathbf{-5} \\ 0 & 1 & \bigm| & 1 \end{bmatrix}$$

$$16(0) + 1 = 1 \qquad 16(1) + (-16) = \mathbf{0} \qquad 16(1) + (-21) = -5 \qquad \text{This is the solution matrix.}$$

(f) The solution set: $x = -5$, $y = 1$. As an ordered pair, the solution is $(-5, 1)$.

Solving Systems of Three Linear Equations in Three Variables

Using similar procedures, the Gauss-Jordan Method may be used to solve system of three linear equations in three variables.

Given a system of linear equations

$$Ax + By + Cz = D$$
$$Ex + Fy + Gz = H$$
$$Jx + Ky + Lz = M,$$

$$\downarrow$$

set up the augmented matrix:

$$\begin{bmatrix} A & B & C & | & D \\ E & F & G & | & H \\ J & K & L & | & M \end{bmatrix}$$

Use the row operations to get $\downarrow$
the **solution matrix**: $\downarrow$

$$\begin{bmatrix} 1 & 0 & 0 & | & p \\ 0 & 1 & 0 & | & q \\ 0 & 0 & 1 & | & r \end{bmatrix}$$

Therefore the solution set is: $x = \mathbf{p}$, $y = \mathbf{q}$, and $z = \mathbf{r}.$

EXAMPLE 7   Solve by using the Gauss–Jordan Method:

$$2x + y + 2z = 9$$
$$x - y - z = 2$$
$$x + y - 3z = -4$$

Answer:

(a) Set up the augmented matrix:

$$\begin{bmatrix} 2 & 1 & 2 & | & 9 \\ 1 & -1 & -1 & | & 2 \\ 1 & 1 & -3 & | & -4 \end{bmatrix}$$

(b) Use matrix (a) to get **1** in the first position of row 1. Interchange (switch) row 1 and row 2:

$$R_1 \longleftrightarrow R_2 \qquad \begin{bmatrix} 2 & 1 & 2 & | & 9 \\ 1 & -1 & -1 & | & 2 \\ 1 & 1 & -3 & | & -4 \end{bmatrix} \rightarrow \begin{bmatrix} \mathbf{1} & -1 & -1 & | & 2 \\ 2 & 1 & 2 & | & 9 \\ 1 & 1 & -3 & | & -4 \end{bmatrix}$$

(c) Use matrix (b): Leave row 1 unchanged.
To get **0** in the first position of row 2, multiply –2 times row 1 and add to row 2; write the result on row 2. To get **0** in the first position row 3, subtract row 3 from row 1; write the result on row 3.

$$\begin{array}{c} -2R_1 + R_2 \rightarrow R_2 \\ R_1 - R_3 \rightarrow R_3 \end{array} \qquad \begin{bmatrix} 1 & -1 & -1 & | & 2 \\ 2 & 1 & 2 & | & 9 \\ 1 & 1 & -3 & | & -4 \end{bmatrix} \rightarrow \begin{bmatrix} 1 & -1 & -1 & | & 2 \\ \mathbf{0} & 3 & 4 & | & 5 \\ \mathbf{0} & -2 & 2 & | & 6 \end{bmatrix}$$

181

(8.5)

(d) Use matrix (c) to get **1** in the second position of row 2. Add row 3 to row 2 and write the result on row 2:

$$R_3 + R_2 \rightarrow R_2 \qquad \begin{bmatrix} 1 & -1 & -1 & | & 2 \\ 0 & 3 & 4 & | & 5 \\ 0 & -2 & 2 & | & 6 \end{bmatrix} \rightarrow \begin{bmatrix} 1 & -1 & -1 & | & 2 \\ \mathbf{0} & \mathbf{1} & \mathbf{6} & | & \mathbf{11} \\ 0 & -2 & 2 & | & 6 \end{bmatrix}$$

(e) Use matrix (d): Leave row 2 unchanged.
To get **0** in the second position row 1, add row 2 to row 1; write the result on row 1. To get **0** in the second position of row 3, multiply 2 times row 2 and add to row 3; write the result on row 3.

$$R_2 + R_1 \rightarrow R_1$$
$$2R_2 + R_3 \rightarrow R_3 \qquad \begin{bmatrix} 1 & -1 & -1 & | & 2 \\ 0 & 1 & 6 & | & 11 \\ 0 & -2 & 2 & | & 6 \end{bmatrix} \rightarrow \begin{bmatrix} \mathbf{1} & \mathbf{0} & \mathbf{5} & | & \mathbf{13} \\ 0 & 1 & 6 & | & 11 \\ \mathbf{0} & \mathbf{0} & \mathbf{14} & | & \mathbf{28} \end{bmatrix}$$

(f) Use matrix (e) to get **1** in the third position of row 3. Divide row 3 by 14 and write the result on row 3:

$$\frac{R_3}{14} \rightarrow R_2 \qquad \begin{bmatrix} 1 & 0 & 5 & | & 13 \\ 0 & 1 & 6 & | & 11 \\ 0 & 0 & 14 & | & 28 \end{bmatrix} \rightarrow \begin{bmatrix} 1 & 0 & 5 & | & 13 \\ 0 & 1 & 6 & | & 11 \\ \mathbf{0} & \mathbf{0} & \mathbf{1} & | & \mathbf{2} \end{bmatrix}$$

(g) Use matrix (f): Leave row 3 unchanged.
To get **0** in the third position row 1, multiply – 5 times row 3 and add to row 1; write the result on row 1. To get **0** in the third position of row 2, multiply – 6 times row 3 and add to row 2; write the result on row 2.

$$-5R_3 + R_1 \rightarrow R_1$$
$$-6R_3 + R_2 \rightarrow R_2 \qquad \begin{bmatrix} 1 & 0 & 5 & | & 13 \\ 0 & 1 & 6 & | & 11 \\ 0 & 0 & 1 & | & 2 \end{bmatrix} \rightarrow \begin{bmatrix} \mathbf{1} & \mathbf{0} & \mathbf{0} & | & \mathbf{3} \\ \mathbf{0} & \mathbf{1} & \mathbf{0} & | & \mathbf{-1} \\ 0 & 0 & 1 & | & 2 \end{bmatrix}$$

This is the solution matrix.

(h) The solution set: x = 3, y = – 1, z = 2. As an ordered triple, the solution is (3, – 1, 2).

SOLVING INCONSISTENT AND DEPENDENT SYSTEMS

The following hints and examples will show how to use the Gauss-Jordan Matrix Method to recognize and solve systems of linear equations which are inconsistent or dependent.

Solving Inconsistent Systems

✠ *__Helpful Hints__* – When using the row operations to transform an augmented matrix:
(1) If a row has all zeros to left of the vertical line and a number other than zero to the right of the line, then the system of equations is an **inconsistent system**. The system has *no solution*. (2) These rules apply to inconsistent systems of *any size*. Refer to the next example.

(8.5)

EXAMPLE 8 Solve by using the Gauss–Jordan Method: $x - y = 5$
$-2x + 2y = -1$

Answer:

(a) Set up the augmented matrix: $\left[\begin{array}{cc|c} 1 & -1 & 5 \\ -2 & 2 & -1 \end{array}\right]$

(b) The first position of row 1 is already **1**. Use matrix (a) to get **0** in the first position of row 2. Multiply 2 times row 1 and add to row 2; write the result on row 2:

$$2R_1 + R_2 \to R_2 \qquad \left[\begin{array}{cc|c} 1 & -1 & 5 \\ -2 & 2 & -1 \end{array}\right] \to \left[\begin{array}{cc|c} 1 & -1 & 5 \\ \mathbf{0} & \mathbf{0} & \mathbf{9} \end{array}\right]$$

(c) The second row of the augmented matrix has *zeros on the left side of the vertical line and 9 on the right side of the line.* The second row corresponds to the equation $0 = 9$ which is FALSE. Therefore this is an **inconsistent system**. This system has ***No Solution***.

Solving Dependent Systems

Helpful Hints – When using the row operations to transform an augmented matrix:
(1) If an entire row is all zeros, then the system of equations is a **dependent system**. The system has *infinite solutions*. (2) These rules apply to dependent systems of *any size*.

EXAMPLE 9 Solve by using the Gauss–Jordan Method:
$x - y + z = 1$
$x + y - z = 3$
$x + 3y - 3z = 5$

Answer:

(a) Set up the augmented matrix: $\left[\begin{array}{ccc|c} 1 & -1 & 1 & 1 \\ 1 & 1 & -1 & 3 \\ 1 & 3 & -3 & 5 \end{array}\right]$

(b) The first position of row 1 is already **1**. Use matrix (a): Leave row 1 unchanged.
To get **0** in the first position of row 2, subtract row 2 from row 1; write the result on row 2. To get **0** in the first position row 3, subtract row 3 from row 1; write the result on row 3.

$$\begin{array}{l} R_1 - R_2 \to R_2 \\ R_1 - R_3 \to R_3 \end{array} \qquad \left[\begin{array}{ccc|c} 1 & -1 & 1 & 1 \\ 1 & 1 & -1 & 3 \\ 1 & 3 & -3 & 5 \end{array}\right] \to \left[\begin{array}{ccc|c} 1 & -1 & 1 & 1 \\ \mathbf{0} & \mathbf{-2} & \mathbf{2} & \mathbf{-2} \\ \mathbf{0} & \mathbf{-4} & \mathbf{4} & \mathbf{-4} \end{array}\right]$$

(c) Use matrix (b) to get **1** in the second position of row 2. Divide row 2 by – 2 and write the result on row 2:

$$\frac{R_2}{-2} \to R_2 \qquad \left[\begin{array}{ccc|c} 1 & -1 & 1 & 1 \\ 0 & -2 & 2 & -2 \\ 0 & -4 & 4 & -4 \end{array}\right] \to \left[\begin{array}{ccc|c} 1 & -1 & 1 & 1 \\ \mathbf{0} & \mathbf{1} & \mathbf{-1} & \mathbf{1} \\ 0 & -4 & 4 & -4 \end{array}\right]$$

(8.5)

(d) Use matrix (c): Leave row 2 unchanged.
To get **0** in the second position row 1, add row 2 to row 1; write the result on row 1. To get **0** in the
second position of row 3, multiply 4 times row 2 and add to row 3; write the result on row 3.

$$R_2 + R_1 \rightarrow R_1$$

$$4R_2 + R_3 \rightarrow R_3$$

$$\left[\begin{array}{ccc|c} 1 & -1 & 1 & 1 \\ 0 & 1 & -1 & 1 \\ 0 & -4 & 4 & -4 \end{array}\right] \rightarrow \left[\begin{array}{ccc|c} \mathbf{1} & \mathbf{0} & \mathbf{0} & \mathbf{2} \\ 0 & 1 & -1 & 1 \\ \mathbf{0} & \mathbf{0} & \mathbf{0} & \mathbf{0} \end{array}\right]$$

(h) The third row of the augmented matrix has *all zeros*. The third row corresponds to the equation
$0 = 0$ which is TRUE. Therefore this is a **dependent system**. This system has ***infinite solutions***.

Write the corresponding equations for row 1 and row 2.

$$x = 2$$
$$y - z = 1$$

Solve the second equation for y.

$$y = z + 1$$

Let z be any real number.

$$z = \text{any real number}$$

Solution set: The set of all ordered triples $\{(x, y, z) \mid (x, y, z) = (2, z + 1, z)\}$ where z is any real
number.

8.6   DETERMINANTS AND CRAMER'S RULE

DEFINITION OF DETERMINANT

Recall that a square matrix is an $n \times n$ matrix. It has the same number of rows and columns. The
following are square matrices:

$$\begin{bmatrix} 2 & 4 \\ 3 & -1 \end{bmatrix}$$

2 x 2 matrix

$$\begin{bmatrix} 0 & -6 & 7 \\ 2 & 3 & -9 \\ -1 & 1 & -2 \end{bmatrix}$$

3 x 3 matrix

Each square matrix is associated with a unique number called the **determinant**.   The symbol for
determinant is *vertical lines around a matrix instead of brackets*. The determinant answer is a *number* (with no
lines or brackets!).

The symbol $\begin{vmatrix} 2 & 4 \\ 3 & -1 \end{vmatrix}$ means find the determinant for the matrix $\begin{bmatrix} 2 & 4 \\ 3 & -1 \end{bmatrix}$.

The symbol $\begin{vmatrix} 0 & -6 & 7 \\ 2 & 3 & -9 \\ -1 & 1 & -2 \end{vmatrix}$ means find the determinant for the matrix $\begin{bmatrix} 0 & -6 & 7 \\ 2 & 3 & -9 \\ -1 & 1 & -2 \end{bmatrix}$.

DETERMINANT OF A 2 x 2 MATRIX

> To find the determinant for the 2 x 2 matrix $\begin{bmatrix} a & b \\ c & d \end{bmatrix}$,
>
> cross-multiply and subtract: $\begin{vmatrix} a & b \\ c & d \end{vmatrix} = ad - bc$
>
> The answer to the determinant is a number.

EXAMPLE 1   Evaluate each determinant:

(a) $\begin{vmatrix} 2 & 4 \\ 3 & -1 \end{vmatrix}$ (b) $\begin{vmatrix} 5 & -4 \\ 3 & -2 \end{vmatrix}$

Answers:

(a) $\begin{vmatrix} 2 & 4 \\ 3 & -1 \end{vmatrix} = (2)(-1) - (3)(4)$

$= -2 - 12$

$= -14$

(b) $\begin{vmatrix} 5 & -4 \\ 3 & -2 \end{vmatrix} = (5)(-2) - (3)(-4)$

$= -10 - (-12) = -10 + 12$

$= 2$

DETERMINANT OF A 3 x 3 MATRIX

For the determinant $\begin{vmatrix} A_1 & B_1 & C_2 \\ A_2 & B_2 & C_2 \\ A_3 & B_3 & C_3 \end{vmatrix}$,

- The **minor determinant for A_1** is found by deleting the row and the column which contain A_1: $\begin{vmatrix} B_2 & C_2 \\ B_3 & C_3 \end{vmatrix}$

- The **minor determinant for A_2** is found by deleting the row and the column which contain A_2: $\begin{vmatrix} B_1 & C_1 \\ B_3 & C_3 \end{vmatrix}$

- The **minor determinant for A_3** is found by deleting the row and the column which contain A_3: $\begin{vmatrix} B_1 & C_1 \\ B_2 & C_2 \end{vmatrix}$

> ## DETERMINANT OF A 3 x 3 MATRIX:
>
> To find the determinant of a the 3 x 3 matrix, set up the **expansion by minors**:
>
> $$\begin{vmatrix} A_1 & B_1 & C_2 \\ A_2 & B_2 & C_2 \\ A_3 & B_3 & C_3 \end{vmatrix} = A_1 \begin{vmatrix} B_2 & C_2 \\ B_3 & C_3 \end{vmatrix} - A_2 \begin{vmatrix} B_1 & C_1 \\ B_3 & C_3 \end{vmatrix} + A_3 \begin{vmatrix} B_1 & C_1 \\ B_2 & C_2 \end{vmatrix}$$
>
> Then simplify to get a single number.

185

(8.6)

Thus $\begin{vmatrix} A_1 & 4 & 7 \\ A_2 & 5 & 8 \\ A_3 & 6 & 9 \end{vmatrix} = A_1 \begin{vmatrix} 5 & 8 \\ 6 & 9 \end{vmatrix} - A_2 \begin{vmatrix} 4 & 7 \\ 6 & 9 \end{vmatrix} + A_3 \begin{vmatrix} 4 & 7 \\ 5 & 8 \end{vmatrix}$

$$= A_1(45 - 48) - A_2(36 - 42) + A_3(32 - 35) \ldots \text{etc.}$$

EXAMPLE 2   Evaluate $\begin{vmatrix} 0 & -6 & 7 \\ 2 & 3 & -9 \\ -1 & 1 & -2 \end{vmatrix}$ using expansion by minors for the first column:

Answer:

$$\begin{vmatrix} 0 & -6 & 7 \\ 2 & 3 & -9 \\ -1 & 1 & -2 \end{vmatrix} = 0 \begin{vmatrix} 3 & -9 \\ 1 & -2 \end{vmatrix} - 2 \begin{vmatrix} -6 & 7 \\ 1 & -2 \end{vmatrix} + (-1) \begin{vmatrix} -6 & 7 \\ 3 & -9 \end{vmatrix}$$

$$= 0 [-6 - (-9)] - 2 (12 - 7) - 1 (54 - 21)$$

$$= 0(3) - 2(5) - 1(33) = 0 - 10 - 33 = -43$$

The determinant is -43.

✠ *__Helpful Hints__* - (1) When setting up the expansion by minors for the 1st column, 3rd column, 1st row or 3rd row, the signs for the three terms are: positive, negative, positive. (2) When setting up the expansion by minors for the 2nd row or 2nd column, the signs are: negative, positive, negative. (3) It is easier to find the determinant if a row or column contains zero.

EXAMPLE 3   Evaluate $\begin{vmatrix} 0 & -6 & 7 \\ 2 & 3 & -9 \\ -1 & 1 & -2 \end{vmatrix}$ using expansion by minors for the second row.

Answer:

$$\begin{vmatrix} 0 & -6 & 7 \\ 2 & 3 & -9 \\ -1 & 1 & -2 \end{vmatrix} = (-)2 \begin{vmatrix} -6 & 7 \\ 1 & -2 \end{vmatrix} + 3 \begin{vmatrix} 0 & 7 \\ -1 & -2 \end{vmatrix} - (-9) \begin{vmatrix} 0 & -6 \\ -1 & 1 \end{vmatrix}$$

$$= -2 (12 - 7) + 3[0 - (-7)] + 9(0 - 6)$$

$$= -2 (5) + 3(7) + 9 (-6) = -10 + 21 - 54 = -43$$

The determinant is -43.

✠ *__Helpful Hint__* – The previous examples show that when evaluating a 3 x 3 determinant, you may choose any row or column (and the appropriate signs); the result will be the same answer.

For the remaining examples of this chapter, the Study Guide will show the determinant for a 3 x 3 matrix using expansion by minors for the first column.

CRAMER'S RULE

Using Determinants to Solve Linear Systems in Two Variables

Given the linear system $A_1X + B_1Y = C_1$, the **coefficient matrix** is $\begin{bmatrix} A_1 & B_1 \\ A_2 & B_2 \end{bmatrix}$.
$A_2X + B_2Y = C_2$

The coefficients of x are in the first column and the coefficients of y are in the second column.

Let $\mathbf{D} = \begin{vmatrix} A_1 & B_1 \\ A_2 & B_2 \end{vmatrix}$, $\mathbf{D_X} = \begin{vmatrix} C_1 & B_1 \\ C_2 & B_2 \end{vmatrix}$, and $\mathbf{D_Y} = \begin{vmatrix} A_1 & C_1 \\ A_2 & C_2 \end{vmatrix}$.

D is the determinant of the coefficient matrix. The determinant D_x is set up by replacing the first column (x column) of D with the constants C_1 and C_2. The determinant D_y is set up by replacing the second column (y column) of D with the constants C_1 and C_2.

Cramer's Rule   (For Systems of Two Linear Equations in Two Variables)

Given the linear system $A_1X + B_1Y = C_1$
$A_2X + B_2Y = C_2$

let $\mathbf{D} = \begin{vmatrix} A_1 & B_1 \\ A_2 & B_2 \end{vmatrix}$, $D_X = \begin{vmatrix} C_1 & B_1 \\ C_2 & B_2 \end{vmatrix}$, and $D_Y = \begin{vmatrix} A_1 & C_1 \\ A_2 & C_2 \end{vmatrix}$.

Then $\mathbf{X} = \dfrac{D_X}{D}$ **and** $\mathbf{Y} = \dfrac{D_Y}{D}$ where $D \neq 0$.

EXAMPLE 4   Use Cramer's Rule to solve the system: $3x + 2y = -8$
$9x + y = 1$

Answer:

Evaluate three determinants.

$D = \begin{vmatrix} 3 & 2 \\ 9 & 1 \end{vmatrix} = (3)(1) - (9)(2) = 3 - 18 = -15$

$D_x = \begin{vmatrix} -8 & 2 \\ 1 & 1 \end{vmatrix} = (-8)(1) - (1)(2) = -8 - 2 = -10$ $\qquad D_y = \begin{vmatrix} 3 & -8 \\ 9 & 1 \end{vmatrix} = 3 - (-72) = 75$

By Cramer's Rule, $x = \dfrac{D_x}{D} = \dfrac{-10}{-15} = \dfrac{2}{3}$ and $y = \dfrac{D_y}{D} = \dfrac{75}{-15} = -5$.

The solution set is the ordered pair $\left(\dfrac{2}{3}, -5 \right)$.

(8.6)

Using Determinants to Solve Linear Systems in Three Variables

Cramer's Rule   (For Systems of Three Linear Equations in Three Variables)

Given the linear system
$$\begin{array}{ccccccc} a_1x & + & b_1y & + & c_1z & = & d_1 \\ a_2x & + & b_2y & + & c_2z & = & d_2 \\ a_3x & + & b_3y & + & c_3z & = & d_3 \end{array}, \quad \text{let } D = \begin{vmatrix} a_1 & b_1 & c_2 \\ a_2 & b_2 & c_2 \\ a_3 & b_3 & c_3 \end{vmatrix}$$

let $D_X = \begin{vmatrix} d_1 & b_1 & c_2 \\ d_2 & b_2 & c_2 \\ d_3 & b_3 & c_3 \end{vmatrix}$, $D_Y = \begin{vmatrix} a_1 & d_1 & c_2 \\ a_2 & d_2 & c_2 \\ a_3 & d_3 & c_3 \end{vmatrix}$, and $D_Z = \begin{vmatrix} a_1 & b_1 & d_2 \\ a_2 & b_2 & d_2 \\ a_3 & b_3 & d_3 \end{vmatrix}$.

Then $x = \dfrac{D_X}{D}$, $y = \dfrac{D_Y}{D}$, and $z = \dfrac{D_Z}{D}$ where $D \neq 0$.

EXAMPLE 5   Use Cramer's Rule to solve the system:
$$\begin{array}{ccccc} x & & & -z & = -1 \\ & y & & -z & = -2 \\ x & + & y & & = -2 \end{array}$$

Answer:

Use expansion by minors for the first column to find the determinants D, D_x, D_y, and D_z.

$$D = \begin{vmatrix} 1 & 0 & -1 \\ 0 & 1 & -1 \\ 1 & 1 & 0 \end{vmatrix} = 1\begin{vmatrix} 1 & -1 \\ 1 & 0 \end{vmatrix} - 0\begin{vmatrix} 0 & -1 \\ 1 & 0 \end{vmatrix} + 1\begin{vmatrix} 0 & -1 \\ 1 & -1 \end{vmatrix}$$

$$= 1[0-(-1)] - 0 + 1[0-(-1)]$$

$$= 1(1) - 0 + 1(1) = 1 - 0 + 1 = 2$$

$$D_x = \begin{vmatrix} -1 & 0 & -1 \\ -2 & 1 & -1 \\ -2 & 1 & 0 \end{vmatrix} = (-1)\begin{vmatrix} 1 & -1 \\ 1 & 0 \end{vmatrix} - (-2)\begin{vmatrix} 0 & -1 \\ 1 & 0 \end{vmatrix} + (-2)\begin{vmatrix} 0 & -1 \\ 1 & -1 \end{vmatrix}$$

$$= -1[0-(-1)] + 2[0-(-1)] - 2[0-(-1)]$$

$$= -1(1) + 2(1) - 2(1) = -1 + 2 - 2 = -1$$

$$D_y = \begin{vmatrix} 1 & -1 & -1 \\ 0 & -2 & -1 \\ 1 & -2 & 0 \end{vmatrix} = 1\begin{vmatrix} -2 & -1 \\ -2 & 0 \end{vmatrix} - 0\begin{vmatrix} -1 & -1 \\ -2 & 0 \end{vmatrix} + 1\begin{vmatrix} -1 & -1 \\ -2 & -1 \end{vmatrix}$$

$$= 1(0-2) - 0 + 1(1-2)$$

$$= 1(-2) - 0 + 1(-1) = -2-0-1 = -3$$

$$D_z = \begin{vmatrix} 1 & 0 & -1 \\ 0 & 1 & -2 \\ 1 & 1 & -2 \end{vmatrix} = 1\begin{vmatrix} 1 & -2 \\ 1 & -2 \end{vmatrix} - 0\begin{vmatrix} 0 & -1 \\ 1 & -2 \end{vmatrix} + 1\begin{vmatrix} 0 & -1 \\ 1 & -2 \end{vmatrix}$$

$$= 1[-2-(-2)] - 0 + 1[0-(-1)]$$

$$= 1(0) - 0 + 1(1) = 0 - 0 + 1 = 1$$

By Cramer's Rule, $x = \dfrac{D_X}{D} = \dfrac{-1}{2}$, $\quad y = \dfrac{D_Y}{D} = \dfrac{-3}{2}$ $\quad$ and $\quad z = \dfrac{D_Z}{D} = \dfrac{1}{2}$.

The solution set is the ordered triple $\left(-\dfrac{1}{2}, -\dfrac{3}{2}, \dfrac{1}{2} \right)$.

SOLVING INCONSISTENT AND DEPENDENT SYSTEMS

IF THE DETERMINANT **D = 0**:

Cramer's Rule does not apply. The system is either inconsistent and has no solution, or, the system is dependent and has infinite solutions. *Use the Addition Method* to find the answer.

EXAMPLE 6   Solve the system: $\quad 4x - 2y = -5$
$\qquad\qquad\qquad\qquad\qquad\qquad -6x + 3y = 2$

Answer:

The determinant $D = \begin{vmatrix} 4 & -2 \\ -6 & 3 \end{vmatrix} = 12 - 12 = \underline{0}$. Cramer's Rule does not apply.

Using the Addition Method,

Multiply the first equation times 3 and multiply the second $\qquad\qquad$ $3(4x - 2y = -5)$
Equation times 2. $\qquad\qquad\qquad\qquad\qquad\qquad\qquad\qquad\qquad$ $2(-6x + 3y = 2)$

Rewrite the system. $\qquad\qquad\qquad\qquad\qquad\qquad\qquad\qquad\qquad$ $12x - 6y = -15$
$\qquad\qquad\qquad\qquad\qquad\qquad\qquad\qquad\qquad\qquad\qquad$ $\underline{-12x + 6y = \quad 4}$

Add both equations together. $\qquad\qquad\qquad\qquad\qquad\qquad\qquad\qquad$ $0 = -11$

Both variables canceled out and $0 \neq -11$. $\qquad\qquad\qquad\qquad\qquad$ This is FALSE.

This system has *No Solution*.

(8.6)

EXAMPLE 7   Solve the system:

$$x + y - 4z = 6$$
$$0.5x - 0.5y + z = 2$$
$$-x + y - 2z = -4$$

Answer:

Use expansion by minors for the first column to find the determinant D.

$$D = \begin{vmatrix} 1 & 1 & -4 \\ 0.5 & -0.5 & 1 \\ -1 & 1 & -2 \end{vmatrix} = 1\begin{vmatrix} -0.5 & 1 \\ 1 & -2 \end{vmatrix} - 0.5\begin{vmatrix} 1 & -4 \\ 1 & -2 \end{vmatrix} + (-1)\begin{vmatrix} 1 & -4 \\ -0.5 & 1 \end{vmatrix}$$

$$= 1(1 - 1) - 0.5[-2 - (-4)] - 1(1 - 2)$$
$$= 1(0) - 0.5(2) - 1(-1) = 0 - 1 + 1 = \underline{0}$$

The determinant D = 0; Cramer's Rule does not apply. Use the Addition Method to solve the system.

Number the equations.

[E1] $x + y - 4z = 6$
[E2] $0.5x - 0.5y + z = 2$
[E3] $-x + y - 2z = -4$

For equations [E1] and [E2], use the Addition procedure to cancel out the x-column. The result is equation [E4].

$$\begin{array}{r} x + y - 4z = 6 \\ -2(0.5x - 0.5y + z = 2) \end{array}$$

$$\begin{array}{r} x + y - 4z = 6 \\ -x + y - 2z = -4 \\ \hline \end{array}$$
[E4] $2y - 6z = 2$

For equations [E1] and [E3], use the Addition procedure to cancel out the x-column. The result is equation [E5].

$$\begin{array}{r} x + y - 4z = 6 \\ -x + y - 2z = -4 \\ \hline \end{array}$$

Observe that equations [E4] and [E5] are the same.

[E5] $2y - 6z = 2$

Multiply – 1 times equation [E5] and add both equations together.

[E4] $2y - 6z = 2$
$-1 \cdot$ [E5] $\underline{-2y + 6z = -2}$
$0 = 0$

Both variables canceled out and the result is a true statement.   This is TRUE.
This is a **dependent system**. It has *infinite solutions*.

Let z be any real number. z = any real number

Solve equation [E4] for y.

[E4] $2y - 6z = 2$
$2y = 6z + 2$
$y = 3z + 1$

Using equation [E1], substitute $3z + 1$ for y.

[E1] $x + y - 4z = 6$
$x + (3z + 1) - 4z = 6$

Solve the equation for x.

$x - z + 1 = 6$
$x = z + 5$

Solution Set: The set of all ordered triples $\{(x, y, z) \mid (x, y, z) = (z + 5, 3z + 1, z)\}$ where z is any real number.

Chapter 9 Factorial Notation and Applications

9.1 FACTORIAL NOTATION

The Math symbol **n!** is read, "n factorial" where n is a positive integer. The expression *n!* means multiply every whole number *from n down to 1*. Therefore,

> **n Factorial**
> For any positive integer n, **n!** = $n(n-1)(n-2)\cdot\cdot\cdot 3\cdot 2\cdot 1$

EXAMPLE 1 Evaluate the factorials:

(a) $3! = 3\cdot 2\cdot 1 = 6$ (b) $5! = 5\cdot 4\cdot 3\cdot 2\cdot 1 = 120$

Helpful Hints – **Factorial Notation and the Calculator:** (1) For operations written above a calculator key, the *Mathco* study guide will show the [2nd] key; press the key which is on *your* calculator, [2nd] or [Shift] or [INV]. (2) On the scientific calculator, the factorial symbol is usually shown as [2nd] [x!].

EXAMPLE 2 Use a scientific calculator to evaluate each factorial.

(a) 5!	Calculator:	5 [2nd] [x!]	Answer: 120
(a) 8!	Calculator:	8 [2nd] [x!]	Answer: 40,320

> **Special rules for n!**
> - $0! = 1$
> - $1! = 1$
> - $n! = n(n-1)!$ or $n! = n(n-1)(n-2)!$
>   or $n! = n(n-1)(n-2)(n-3)!$ etc.

EXAMPLE 3 Using the definition of factorial:

$8! = 8\cdot 7\cdot 6\cdot 5\cdot 4\cdot 3\cdot 2\cdot 1$, $7! = 7\cdot 6\cdot 5\cdot 4\cdot 3\cdot 2\cdot 1$ and $6! = 6\cdot 5\cdot 4\cdot 3\cdot 2\cdot 1$.

Observe that 7! is part of 8!; 6! is also part of 8!. Therefore 8! can be written as:

$8! = 8\cdot 7!$ or $8! = 8\cdot 7\cdot 6!$ or $8! = 8\cdot 7\cdot 6\cdot 5!$ etc.

As you expand a factorial, you may stop the product at any lower factorial.

EXAMPLE 4 Rewrite and evaluate each of the following expressions. Use the Special Rules for n! as shortcuts for the fraction problems. Or, use a scientific calculator.

(Rewritten) (Calculator Keys)

(a) $\frac{5!}{4!} = \frac{5\cdot 4!}{4!} = \frac{5\cdot \cancel{4!}}{\cancel{4!}} = 5$ 5 [2nd] [x!] [÷] 4 [2nd] [x!] [=]

(9.1)

(Rewritten)   (Calculator Keys)

(b) $\dfrac{9!}{6!} = \dfrac{9\cdot 8\cdot 7\cdot 6!}{6!} = \dfrac{9\cdot 8\cdot 7\cdot 6!}{6!} = 504$ 9 [2nd] [x!] [÷] 6 [2nd] [x!] [=]

(c) $\dfrac{8!}{5!3!} = \dfrac{8\cdot 7\cdot 6\cdot 5!}{5!\cdot 3\cdot 2\cdot 1} = \dfrac{8\cdot 7\cdot 6\cdot 5!}{5!\cdot 3\cdot 2\cdot 1} = 56$ 8 [2nd] [x!] [÷] 5 [2nd] [x!]
[÷] 3 [2nd] [x!] [=]

(d) $\dfrac{10!}{3!7!} = \dfrac{10\cdot 9\cdot 8\cdot 7!}{3\cdot 2\cdot 1\cdot 7!} = \dfrac{10\cdot \overset{3}{9}\cdot \overset{4}{8}\cdot 7!}{3\cdot 2\cdot 1\cdot 7!} = 120$ 10 [2nd] [x!] [÷] 3 [2nd] [x!]
[÷] 7 [2nd] [x!] [=]

EXAMPLE 5   Evaluate the expression: $\dfrac{400!}{397!}$

Answer:

To evaluate the numerator, use the calculator keys: 400 [2nd] [x!] [=] . The result on the calculator screen is *"Error"* or *"Math Error"* which indicates that the answer to 400! *is too large for the calculator*. Partially expand the numerator and cancel the denominator. Then use a calculator.

$$\frac{400!}{397!} = \frac{400\cdot 399\cdot 398\cdot 397!}{397!} = \frac{400\cdot 399\cdot 398\cdot 397!}{397!} = 400\cdot 399\cdot 398 = 63{,}520{,}800$$

9.2   FUNDAMENTAL COUNTING PRINCIPLE

An **event** is an activity which may have more than one possible result such as tossing a coin, rolling a die, or drawing a card out of a poker deck. An **outcome** is a possible result of an event.

EXAMPLE 1   List the set of possible outcomes for each event: (a) toss a coin   (b) roll a die

Answers:

(a) For tossing a coin, the set of possible outcomes is {heads, tails}.

(b) A die is a cube whose sides are numbered from 1 to 6. For rolling a die, the set of possible outcomes is: {1, 2, 3, 4, 5, 6}.

For two **independent events,** the outcome of one event does not affect the outcome of the other event. "Tossing a coin" and "rolling a die" are two independent events because the result from tossing the coin is not related to the result from rolling the die.

We will show how to use a **tree diagram** to count all the outcomes of a set of two or more events. The total number of outcomes will equal the total number of final branches.

(9.2)

EXAMPLE 2 A local student orchestra has skillfully practiced 2 overtures, 2 sonatas, and 3 piano concertos. The Orchestra Concert at the school's Arts Festival is to consist one overture, one sonata and one piano concerto. Make a tree diagram showing all of the different program arrangements for the concert.

Tree Diagram: (O → Overture, S → Sonata, C → Concerto)

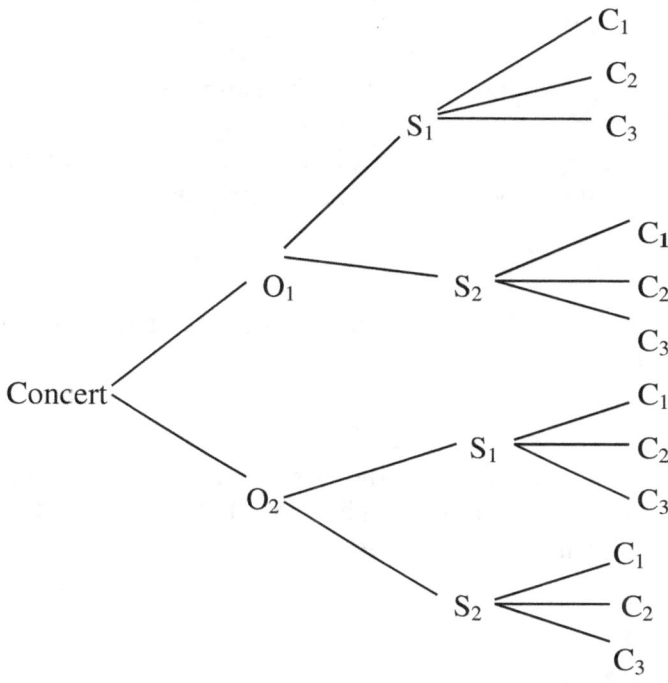

Answer: By counting the final tree branches, there are 12 different program arrangements for the concert.

An alternative method to the tree diagram is the Fundamental Counting Principle.

Fundamental Counting Principle:

Let E_1, E_2, and E_3 represent independent events. If event E_1 has m distinct outcomes, event E_2 has n different outcomes, and event E_3 has p different outcomes, then the **total number of outcomes for events E_1 , E_2 , and E_3 to occur is $m \cdot n \cdot p$.**

EXAMPLE 3 For Example 2 (above), the orchestra conductor can choose from 2 overtures, 2 sonatas and 3 concertos for the concert program. The total number of different concert programs is:

$2 \cdot 2 \cdot 3 = 12$ program arrangements.

(9.2)

EXAMPLE 4 Each license plate of a certain state consists of 3 letters followed by a 3-digit number. How many different license plates are possible if: (a) repetitions of letters and digits are permitted. (b) letters and digits may not be repeated.

Answers:

In the alphabet there are 26 letters. There are 10 single digit numbers from 0 to 9.

(a) Since repetitions are permitted, there are 26 choices for each letter and 10 choices for each digit. The total number of different license plates is $26 \cdot 26 \cdot 26 \cdot 10 \cdot 10 \cdot 10 = 17{,}576{,}000$.

(b) Repetitions are not permitted. The first letter has 26 choices. After choosing the first letter, the second letter has 25 choices remaining. Then the third letter has 24 choices remaining. Similarly, the first digit has 10 choices. After choosing the first digit, the second digit has 9 choices remaining. Then the third digit has 8 choices remaining. The total number of different license plates is $26 \cdot 25 \cdot 24 \cdot 10 \cdot 9 \cdot 8 = 11{,}232{,}000$.

EXAMPLE 5 Find the number of 7-digit local phone numbers which are possible if the first digit cannot be 0, 8, or 9 and the fourth digit cannot be 0.

Answer:

Out of 10 single digit numbers from 0 to 9, there are 7 choices for the first digit, 10 choices for the second digit, 10 choices for the third digit, 9 choices for the fourth digit, and 10 choices each for the remaining three digits. The total number of 7-digit local phone numbers is
$7 \cdot 10 \cdot 10 \cdot 9 \cdot 10 \cdot 10 \cdot 10 = 6{,}300{,}000$.

9.3 PERMUTATIONS AND COMBINATIONS

FINDING THE NUMBER OF PERMUTATIONS

A **permutation** is a rearrangement of objects in which the order of the objects matters. In other words, *a different order is a different answer.* Lotteries, raffle drawings, and the arrangement of books on a shelf are examples of permutations.

EXAMPLE 1 For the Pick-3 Lottery, list the ways the digits 1, 2, 3 can be played.

Answer: 1 2 3   1 3 2   2 1 3   2 3 1   3 1 2   3 2 1

Number of permutations: 6

EXAMPLE 2 Choosing 2 letters at a time, list the ways that the letters h, e, a, d can be written.

Answer: h e   e h   a h   d h

h a   e a   a e   d e

h d   e d   a d   d a

Number of permutations: 12

An alternative method for finding the number of rearrangements of objects is the Permutation Formula.

Permutation Formula

The number of permutations of **n** objects taken **r** at a time is: $P(n, r) = \dfrac{n!}{(n-r)!}$

✳ *Helpful Hint* – The Math symbols for "permutations of n objects taken r at a time" are: $P(n, r)$ or $_nP_r$. Either notation may be used. Set up the formula and use the Special Rules for $n!$ as shortcuts for simplifying the fraction. Or, use a scientific calculator.

EXAMPLE 3 For the Pick-3 Lottery in Example 1, n = 3 digits and r = 3 at a time. Applying the Permutation Formula, the number of ways the digits 1, 2, 3 can be played is:

$$P(3, 3) \;=\; \frac{n!}{(n-r)!} \;=\; \frac{3!}{(3-3)!} \;=\; \frac{3!}{0!} \;=\; \frac{3 \cdot 2 \cdot 1}{1} \;=\; 6$$

EXAMPLE 4 Referring to Example 2, n = 4 letters and r = 2 at a time. The number of ways that the letters h, e, a, d can be written, choosing 2 letters at a time, is:

$$P(4, 2) \;=\; \frac{n!}{(n-r)!} \;=\; \frac{4!}{(4-2)!} \;=\; \frac{4!}{2!} \;=\; \frac{4 \cdot 3 \cdot \cancel{2!}}{\cancel{2!}} \;=\; 12$$

FINDING THE NUMBER OF COMBINATIONS

A **combination** is a grouping of objects in which the order does not matter. In other words, *a different order is the same answer.* Committees and subgroups are examples of combinations.

EXAMPLE 5   List the 3-color pom-poms that can be made for cheerleaders using the colors r = red, b = blue, y = yellow, and w = white.

Answer: For pom-poms, the colors are blended so that you cannot determine the order of the colors. This example is a combination problem. The 3-color pom-poms are:

r b y   r b w   r y w   b y w

Number of color combinations: 4

An alternative method for finding the number of groupings of objects is the Combination Formula.

Combination Formula

The number of combinations of **n** objects taken **r** at a time is: $C(n, r) \;=\; \dfrac{n!}{r!\,(n-r)!}$

✳ *Helpful Hint* – The Math symbols for "combinations of n objects taken r at a time" are: $C(n, r)$ or $_nC_r$. Either notation may be used. Set up the formula and use the Special Rules for $n!$ as shortcuts for simplifying the fraction. Or, use a scientific calculator.

(9.3)

EXAMPLE 6   For the 3-color pom-poms in Example 5, n = 4 colors, r = 3 colors selected at a time. The number of 3-color combinations is:

$$C(4, 3) \; = \; \frac{n!}{r!(n-r)!} = \frac{4!}{3!(4-3)!} = \frac{4!}{3!1!} = \frac{4 \cdot 3!}{3! \cdot 1} = 4$$

EXAMPLE 7   Four identical Fun-Time Watches were purchased for the door prizes at a birthday party. There are 17 children at the party and four children will be selected at random to receive a door prize. In how many ways can four children be selected for the prizes?

Answer:

Since the watches are identical, it does not matter in which order the 4 children are selected. Applying the Combination Formula, n = 17 children and r = 4 chosen at a time. The number of ways 4 children can be selected is:

$$C(17, 4) \; = \; \frac{n!}{r!(n-r)!} = \frac{17!}{4!(17-4)!} = \frac{17!}{4!13!} = \frac{17 \cdot \overset{4}{\cancel{16}} \cdot \overset{5}{\cancel{15}} \cdot \overset{7}{\cancel{14}} \cdot \cancel{13!}}{\cancel{4} \cdot \cancel{3} \cdot \cancel{2} \cdot 1 \cdot \cancel{13!}} = \frac{17 \cdot 4 \cdot 5 \cdot 7}{1} = 2{,}380$$

COMBINATION OR PERMUTATION?

EXAMPLE 8   At the *Neighborhood SummerFest*, Rita, Bill, Yvette, and Wayne signed up to compete in the one-mile race. The 1st place winner will receive a gold medal, 2nd place will receive a silver medal, and 3rd place will receive a bronze medal.

(a) Is this problem describing a combination or permutation?

Answer: This is a permutation problem because it matters who finishes first, second or third.

(b) In how many ways can the 3 medals be awarded to the runners?

Answer: Applying the Permutation Formula, n = 4 runners and r = 3 winners selected at a time. The number of ways the medals can be awarded is:

$$P(4, 3) = \frac{4!}{(4-3)!} = \frac{4!}{1!} = \frac{4 \cdot 3 \cdot 2 \cdot 1}{1} = 24$$

EXAMPLE 9   At a certain college, the Student Senate consists of 15 senators. The college plans to send a delegation of 5 senators to the State Leadership Conference.   According to the guidelines, all 15 senators qualify for the delegation.

(a) Is this problem describing a combination or permutation?

Answer: This is a combination because the order in which the senators are selected for the delegation does not matter.

(b) In how many ways can the delegation of four senators be selected?

Answer: Let n = 15 senators and r = 5 senators chosen at a time. Applying the Combination Formula, the number of ways the delegation can be selected is:

$$C(15, 5) \; = \; \frac{15!}{5!(15-5)!} = \frac{15!}{5!10!} = \frac{\overset{3}{\cancel{15}} \cdot \cancel{14} \cdot 13 \cdot \overset{\cancel{3}}{\cancel{12}} \cdot 11 \cdot \cancel{10!}}{\cancel{5} \cdot \cancel{4} \cdot \cancel{3} \cdot \cancel{2} \cdot 1 \cdot \cancel{10!}} = \frac{3 \cdot 7 \cdot 13 \cdot 11}{1} = 3{,}003$$

9.4   BINOMIAL EXPANSIONS   (THE BINOMIAL THEOREM)

USING POLYNOMIAL MULTIPLICATION

We already know how to use polynomial multiplication to simplify expressions of the form $(x + y)^n$. For instance, $(x + y)^2 = (x + y)(x + y) = x^2 + 2xy + y^2$. To simplify $(x + y)^3$, multiply $(x + y)^2$ times $(x + y)$:

$$(x + y)^3 = (x + y)^2(x + y) = (x^2 + 2xy + y^2)(x + y)$$
$$= x^2(x + y) + 2xy(x + y) + y^2(x + y)$$
$$= x^3 + x^2y + 2x^2y + 2xy^2 + xy^2 + y^3$$
$$= x^3 + 3x^2y + 3xy^2 + y^3$$

The polynomial answer to the expression $(x + y)^n$ is called a **binomial expansion**. The polynomial $x^3 + 3x^2y + 3xy^2 + y^3$ is the binomial expansion for $(x + y)^3$. The following are several binomial expansions.

$$(x + y)^0 = 1$$
$$(x + y)^1 = x + y$$
$$(x + y)^2 = x^2 + 2xy + y^2$$
$$(x + y)^3 = x^3 + 3x^2y + 3xy^2 + y^3$$
$$(x + y)^4 = x^4 + 4x^3y + 6x^2y^2 + 4xy^3 + y^4$$
$$(x + y)^5 = x^5 + 5x^4y + 10x^3y^2 + 10x^2y^3 + 5xy^4 + y^5$$

Given the expression $(x + y)^n$, n is the degree of the binomial. For each binomial expansion we observe that:

- The number of terms in the answer is 1 higher than the degree of the binomial. The answer to $(x + y)^0$ has 1 term, the answer to $(x + y)^1$ has 2 terms, the answer to $(x + y)^2$ has 3 terms, etc.

- The first term of the answer is x^n and the exponents for *x* are decreasing in the polynomial while the exponents for *y* are increasing; the last term is y^n.

- The exponents on the variables for each term of the polynomial add up to the degree of the binomial. For instance, the variables for the terms of $(x + y)^5$ are x^5, x^4y, x^3y^2, x^2y^3, xy^4, and y^5. The exponents of each term add up to 5.

THE BINOMIAL COEFFICIENTS

A different way to find the coefficients of a binomial expansion is to use the **binomial coefficient symbol** $\begin{pmatrix} n \\ r \end{pmatrix}$. It is pronounced "*n* choose *r*." The binomial coefficient symbol $\begin{pmatrix} n \\ r \end{pmatrix}$ is equivalent to the Combination Formula, C(n, r) or $_nC_r$, which was used in the previous section.

(9.4)

To find the coefficients of a binomial expansion we use the formula:

Binomial Coefficients

$$\binom{n}{r} = C(n, r) = {}_nC_r = \frac{n!}{r!(n-r)!} \qquad \text{where } n \text{ and } r \text{ are nonnegative integers and } n > r.$$

As you evaluate $\binom{n}{r}$ you may set up the formula and use the Special Rules for $n!$ as shortcuts for simplifying the fraction. Or, use a scientific calculator.

EXAMPLE 1   Evaluate $\binom{5}{2}$.

Answer: For this example, n = 5 and r = 2.

$$\binom{5}{2} = \frac{5!}{2!(5-2)!} = \frac{5!}{2!3!} = \frac{5 \cdot \cancel{4}^2 \cdot \cancel{3!}}{\cancel{2} \cdot 1 \cdot \cancel{3!}} = \frac{5 \cdot 2}{1} = 10$$

On the scientific calculator: 5 [2nd] [x!]   [÷] 2 [2nd] [x!]   [÷] 3 [2nd] [x!] [=]

✶ *Helpful Hints* – Let n, p, and r be nonnegative integers; also $n > p$, and, $n > r$. Then:

   (a) $\binom{n}{0} = 1$ (b) $\binom{n}{n} = 1$ (c) Whenever $p + r = n$, $\binom{n}{p} = \binom{n}{r}$.

EXAMPLE 2   Show that: (a) $\binom{5}{0} = 1$ (b) $\binom{5}{5} = 1$ (c) $\binom{5}{2} = \binom{5}{3}$ where 2 + 3 = 5.

Answers:

(a) $\binom{5}{0} = \frac{5!}{0!(5-0)!} = \frac{\cancel{5!}}{0!\cancel{5!}} = \frac{1}{1} = 1$

(b) $\binom{5}{5} = \frac{5!}{5!(5-5)!} = \frac{\cancel{5!}}{\cancel{5!}0!} = \frac{1}{1} = 1$

(c) From Example 1 above, $\binom{5}{2} = \frac{5!}{2!3!} = 10$. Also, $\binom{5}{3} = \frac{5!}{3!2!} = \frac{5 \cdot 4 \cdot \cancel{3!}}{\cancel{3!} \cdot 2 \cdot 1} = 10$.

   Observe that 2 + 3 = 5 and $\binom{5}{2} = \binom{5}{3}$.

(9.4)

USING THE BINOMIAL THEOREM

To simplify an expression of the form $(x + y)^n$, an alternative to polynomial multiplication is to use the Binomial Theorem. You may use a calculator to simplify each term of the expansion.

The Binomial Theorem:

$$(x + y)^n = \binom{n}{0} x^n + \binom{n}{1} x^{n-1} y^1 + \binom{n}{2} x^{n-2} y^2 + \ldots + \binom{n}{n-1} x^1 y^{n-1} + \binom{n}{n} y^n$$

where n is a nonnegative integer.

That is:

$(1st + 2nd)^n =$

$$\binom{n}{0}(1st)^n + \binom{n}{1}(1st)^{n-1}(2nd)^1 + \binom{n}{2}(1st)^{n-2}(2nd)^2 + \ldots + \binom{n}{n-1}(1st)^1(2nd)^{n-1} + \binom{n}{n}(2nd)^n$$

EXAMPLE 3   Use the Binomial Theorem to expand and simplify $(x + y)^5$.

Answer: Remember that for each term, the sum of the exponents on the variables is 5.

$$(x + y)^5 = \binom{5}{0} x^5 + \binom{5}{1} x^4 y^1 + \binom{5}{2} x^3 y^2 + \binom{5}{3} x^2 y^3 + \binom{5}{4} x^1 y^4 + \binom{5}{5} y^5$$

$$= \frac{5!}{0!\,5!} x^5 + \frac{5!}{1!\,4!} x^4 y + \frac{5!}{2!\,3!} x^3 y^2 + \frac{5!}{3!\,2!} x^2 y^3 + \frac{5!}{4!\,1!} xy^4 + \frac{5!}{5!\,0!} y^5$$

$$= \frac{5!}{1 \cdot 5!} x^5 + \frac{5 \cdot 4!}{1 \cdot 4!} x^4 y + \frac{5 \cdot 4 \cdot 3!}{2 \cdot 1 \cdot 3!} x^3 y^2 + \frac{5 \cdot 4 \cdot 3!}{3! \cdot 2 \cdot 1} x^2 y^3 + \frac{5 \cdot 4!}{4! \cdot 1} xy^4 + \frac{5!}{5! \cdot 1} y^5$$

$$= x^5 + 5x^4 y + 10x^3 y^2 + 10x^2 y^3 + 5xy^4 + y^5$$

EXAMPLE 4   Use the Binomial Theorem to expand and simplify $(2x - 3y)^3$.

Answer: $$(2x - 3y)^3 = \binom{3}{0}(2x)^3 + \binom{3}{1}(2x)^2(-3y)^1 + \binom{3}{2}(2x)^1(-3y)^2 + \binom{3}{3}(-3y)^3$$

$$= \frac{3!}{0!\,3!}(8x^3) + \frac{3!}{1!\,2!}(4x^2)(-3y) + \frac{3!}{2!\,1!}(2x)(9y^2) + \frac{3!}{3!\,0!}(-27y^3)$$

$$= \frac{3!}{1 \cdot 3!}(8x^3) + \frac{3 \cdot 2!}{1 \cdot 2!}(4x^2)(-3y) + \frac{3 \cdot 2!}{2! \cdot 1}(2x)(9y^2) + \frac{3!}{3! \cdot 1}(-27y^3)$$

$$= 1(8x^3) + 3(-12x^2 y) + 3(18xy^2) + 1(-27y^3)$$

$$= 8x^3 - 36x^2 y + 54xy^2 - 27y^3$$

(9.4)

EXAMPLE 5   Use the Binomial Theorem to show the first four terms of the expansion for $(x + 4)^{10}$.

Answer: (Remember, you may use a calculator to simplify each term of the expansion.)

$$(x + 4)^{10} = \binom{10}{0}(x)^{10} + \binom{10}{1}(x)^9(4)^1 + \binom{10}{2}(x)^8(4)^2 + \binom{10}{3}(x)^7(4)^3 + \cdots$$

$$= \frac{10!}{0!10!}x^{10} + \frac{10!}{1!9!}x^9(4) + \frac{10!}{2!8!}x^8(16) + \frac{10!}{3!7!}x^7(64) + \cdots$$

$$= \frac{\cancel{10!}}{1 \cdot \cancel{10!}}x^{10} + \frac{10 \cdot \cancel{9!}}{1 \cdot \cancel{9!}}(4x^9) + \frac{\overset{5}{\cancel{10}} \cdot 9 \cdot \cancel{8!}}{\cancel{2} \cdot 1 \cdot \cancel{8!}}(16x^8) + \frac{10 \cdot \overset{3}{\cancel{9}} \cdot 8 \cdot \overset{4}{\cancel{7!}}}{\cancel{3} \cdot \cancel{2} \cdot 1 \cdot \cancel{7!}}(64x^7) + \cdots$$

$$= 1 \cdot x^{10} + 10(4x^9) + 45(16x^8) + 120(64x^7) + \cdots$$

$$= x^{10} + 40x^9 + 720x^8 + 7{,}680x^7 + \cdots$$

FINDING A PARTICULAR TERM OF A BINOMIAL EXPANSION

For $(x + y)^n$, to find the **kth term only**:

(1) Let **n** = the exponent of the binomial and **r** = k − 1, the second exponent of the term.

(2) Then the **kth term** is $\binom{n}{r}x^{n-r}y^r$.

EXAMPLE 6   Find the sixth term of $(x + y)^9$.

Answer:

Let n = 9, the exponent of the binomial. Let k = 6, for the sixth term of the expansion.

Let r = 6 − 1 = 5, the second exponent of the sixth term.

Using the formula, the 6th term is: $\binom{9}{5}x^{9-5}y^5 = \frac{9!}{5!4!}x^4y^5 = \frac{9 \cdot \overset{2}{\cancel{8}} \cdot 7 \cdot \overset{\cancel{2}}{\cancel{6}} \cdot \cancel{5!}}{\cancel{5!} \cdot \cancel{4} \cdot \cancel{3} \cdot \cancel{2} \cdot 1}x^4y^5 = 126x^4y^5$

(Remember, you may use a calculator to simplify the expression.)

Appendix   **List of Formulas and Applications**

(Each application is followed by the page number where the concept is explained.)

GEOMETRIC FORMULAS

Rectangle   W

L

Perimeter, *10* $P = 2L + 2W$

Area, *67* $A = LW$

Square

S

Perimeter, *10* $P = 4S$

Area, 67 $A = S^2$

Triangle

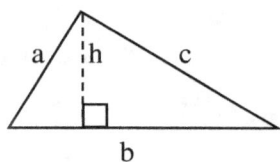

b

Perimeter, *11* $P = a + b + c$

Area, *68* $A = \dfrac{1}{2}bh$

Angles of a Triangle, *9*

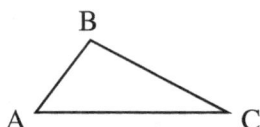

$\angle A + \angle B + \angle C = 180°$

Sides of a Right Triangle, *69*

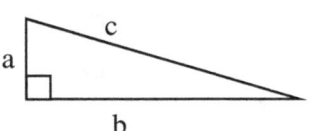

$a^2 + b^2 = c^2$

OTHER APPLICATIONS AND FORMULAS

Binomial Coefficient, *198*

$$\binom{n}{r} = C(n, r) = {}_nC_r = \frac{n!}{r!(n-r)!}$$

Binomial Theorem, *199*

$$(x + y)^n = \binom{n}{0} x^n + \binom{n}{1} x^{n-1} y^1 + \binom{n}{2} x^{n-2} y^2 + \ldots + \binom{n}{n-1} x^1 y^{n-1} + \binom{n}{n} y^n$$

Combination Formula, *195*

$$C(n, r) = {}_nC_r = \frac{n!}{r!(n-r)!}$$

Consecutive Integers, *10*

Cost Function (minimum value), *133*

Distance Formula (see Motion Problems), *14*

$$D = rt$$

Exponential growth, *154*

$$P = P_0 e^{rt}$$

Fundamental Counting Principle, *192*

Interest (Investment) Problems

Simple Interest, *13*

$$I = Prt \qquad (\text{Interest} = \text{principal} \cdot \text{rate} \cdot \text{time})$$

Interest for 1 year = (interest rate)(amount invested)

Compound Interest, *143*

$$A = P\left(1 + \frac{r}{n}\right)^{nt}$$

Continuous Compound Interest, *144*

$$A = Pe^{rt}$$

Present Value, *144*

Lines; Linear Functions

Slope of a line, *38-39*

$$\text{slope } m = \frac{\text{rise}}{\text{run}} = \frac{y_2 - y_1}{x_2 - x_1}$$

Slope of parallel lines, *39*

$$m_2 = m_1$$

Slope of perpendicular lines, *39*

$$m_2 = -\frac{1}{m_1}$$

INDEX

INDEX

~ NOTES ~

~ NOTES ~

www.ingramcontent.com/pod-product-compliance
Lightning Source LLC
Chambersburg PA
CBHW080826220526

45467CB00008B/2205